INTRO TO CHEMISTRY
COLORING WORKBOOK

ISBN: 9781530439799

Table of Contents

Subatomic Particles

Proton   Neutron   Electron

+ Proton has a positive charge
  Neutron has a neutral charge
- Electron has a negative charge

These are the trio of subatomic particles. Choose a color for each and use those colors throughout the book.

Subatomic means: smaller than or occuring within an **atom**.

When these subatomic particles come together, they form atoms. All matter is made of atoms. One proton and one electron form a hydrogen atom.

 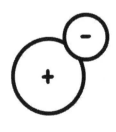

1 proton + 1 electron = a hydrogen atom!

Protons and Elements

Hydrogen is the first element on the periodic table.

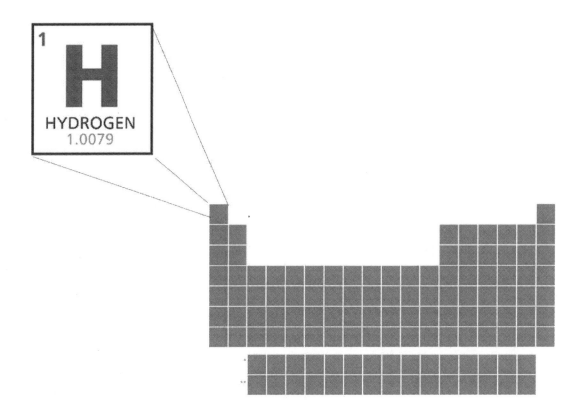

The number of **protons** an atom has tells us the atom's **element**. An element's **atomic number** tells how many protons each atom of that element has.

Every hydrogen atom has only one proton.
Hydrogen's atomic number is one.

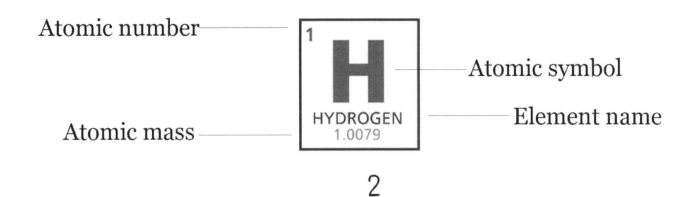

Atomic number

Atomic symbol

Atomic mass

Element name

On the next page is a copy of the periodic table. It is divided into three sections:

1. Metals
2. Nonmetals
3. Metalloids

Metals tend to be good conductors of electricity and heat, and, with mercury as an exception, are solid at room temperature. When subjected to chemical changes, metals lose electrons, while nonmetals gain electrons. Nonmetals are not generally good conductors of heat or electricity, and, depending on the element, can be solid, liquid, or gas at room temperature.

A metalloid is an element that has both metallic and nonmetallic properties. The metalloids sit between the metals and the nonmetals on the periodic table.

Choose a color for each section and color the periodic table. Notice that hydrogen is a nonmetal but is grouped on the left side of the periodic table where the metals are.

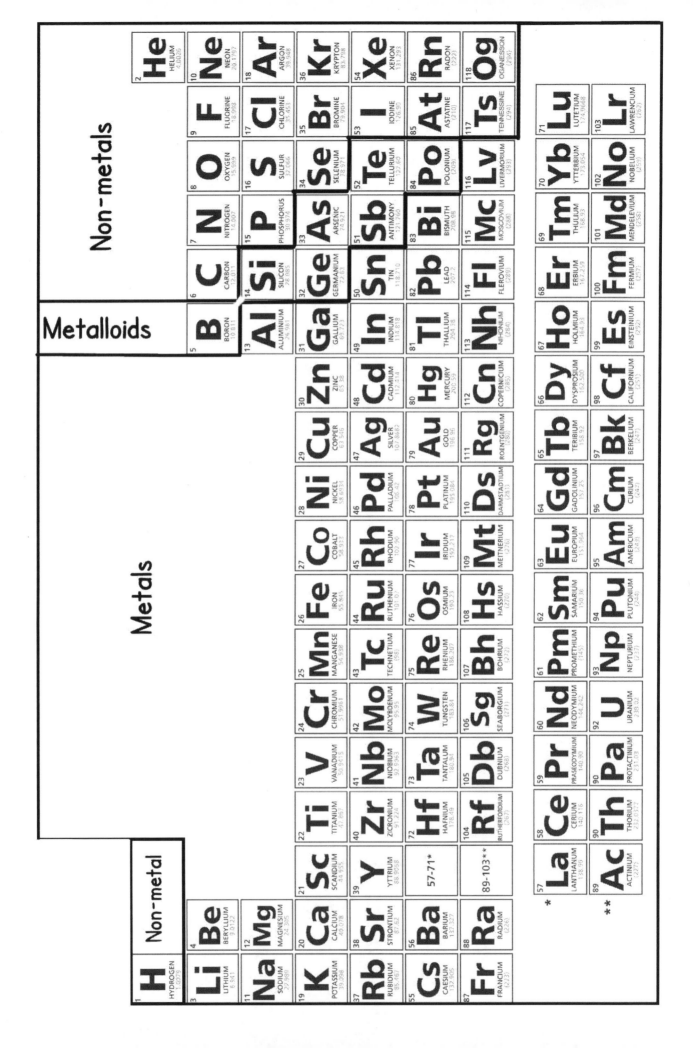

Referring to the periodic table you just colored, tell whether each of these elements is a metal, a nonmetal, or a metalloid. Also write each element's atomic symbol and atomic number:

Hydrogen _____ Symbol: _____ Number: _____

Sodium _____ Symbol: _____ Number: _____

Carbon _____ Symbol: _____ Number: _____

Aluminum _____ Symbol: _____ Number: _____

Potassium _____ Symbol: _____ Number: _____

Oxygen _____ Symbol: _____ Number: _____

Calcium _____ Symbol: _____ Number: _____

Nitrogen _____ Symbol: _____ Number: _____

Boron _____ Symbol: _____ Number: _____

Silicon _____ Symbol: _____ Number: _____

Gold _____ Symbol: _____ Number: _____

Arsenic _____ Symbol: _____ Number: _____

Titanium _____ Symbol: _____ Number: _____

Argon _____ Symbol: _____ Number: _____

Helium _____ Symbol: _____ Number: _____

Electrons and Ions

Atoms have equal amounts of protons and electrons. When the number of protons equals the number of electrons, the atom has a neutral charge. If an atom gains or loses one or more electrons, it is now an **ion** and has either a positive charge or a negative charge.

When an atom is missing an electron, it has a positive charge and is called a **cation.** Hydrogen commonly loses an electron and has a positive charge.

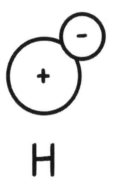

H

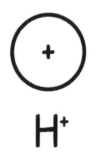

H⁺

A hydrogen atom has:
1 proton
1 electron

neutral charge

A hydrogen ion has:
1 proton
0 electrons

positive charge

When an atom has an extra electron, it has a negative charge and is called an **anion**. Oxygen commonly gains two electrons and has a negative charge.

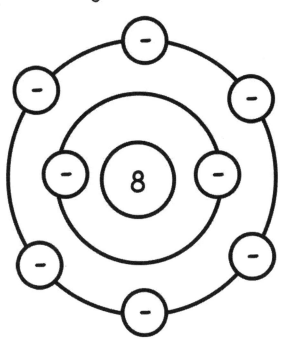

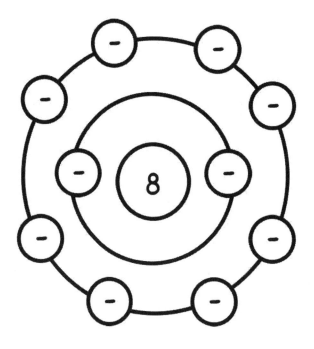

O

O²⁻

An oxygen atom has:

8 protons

8 electrons

neutral charge

An oxygen ion has:

8 protons

10 electrons

negative charge

Electrons **orbit** around the **nucleus** of an atom. The nucleus is where the protons and neutrons are, in the center of the atom.

Color these diagrams. Count the electrons and use the information given to fill in the blanks.

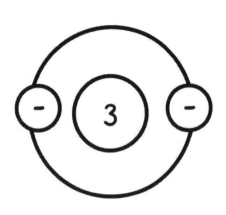

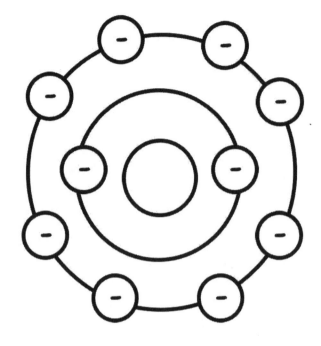

_____

F⁻

A _____ cation has:

A fluorine _____ has:

3 protons
___ electrons

___ protons
___ electrons

_____ charge

_____ charge

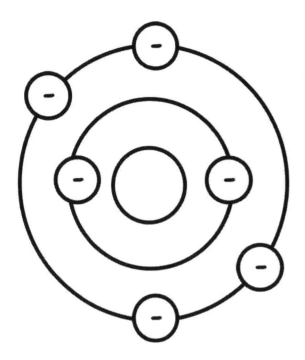

C

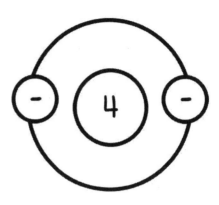

_____

A carbon _____ has:

___ protons
___ electrons

_____ charge

A _____ cation has:

4 protons
___ electrons

_____ charge

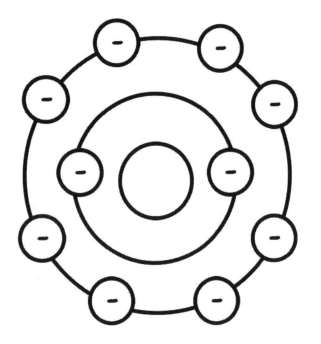

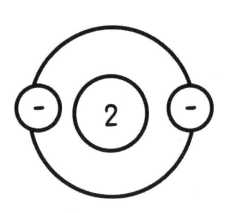

N^{3-}

_____

A nitrogen _____ has:

___ protons
___ electrons

_____ charge

A _____ atom has:

2 protons
___ electrons

_____ charge

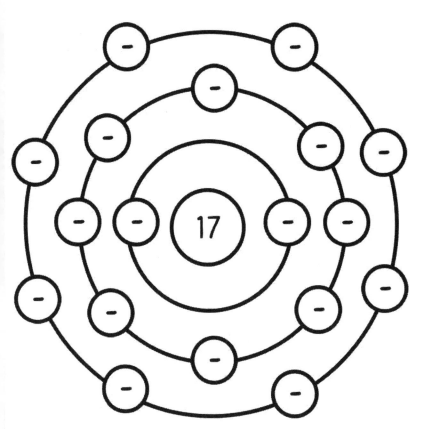

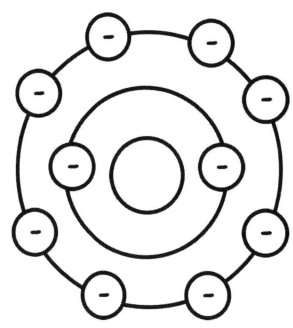

Na⁺

_____

A _____ anion has:

17 protons

___ electrons

_____ charge

A sodium _____ has:

___ protons

___ electrons

_____ charge

11

Neutrons and Isotopes

An atom's nucleus is made up of protons and neutrons. Looking at the periodic table, you can tell how many protons and electrons are in each atom of an element by looking at the atomic number. The number of **neutrons** in each atom, however, can vary, and this variance does not change the atom's charge or alter which element it is.

The number of neutrons an atom has determines that atom's **isotope**.

For example, a carbon atom can have 6, 7, or 8 neutrons. Each of these is a different isotope of carbon. Adding the number of protons to the number of neutrons gives the atom's atomic weight.

An isotope can be indicated with the element's atomic symbol followed by a dash and the isotope's atomic mass. C-12 is an isotope of carbon with an atomic mass of 12.

Fill in the blanks:

C-12 has 6 protons, 6 neutrons, and 6 electrons.

C-13 has 6 protons, ___ neutrons, and 6 electrons.

C-14 has ___ protons, ___ neutrons, and ___ electrons.

The atomic masses listed on the periodic table represent the average weights of all naturally occurring isotopes of that element, weighted in regard to how commonly they occur in nature. One amu, or **atomic mass unit**, is equal to 1/12 the weight of the carbon 12 isotope.

C-12 is by far the most common isotope of carbon, so its atomic mass reflects this by being much closer to 12 than 13 or 14.

 Atomic mass

Using the periodic table and what you know about protons, electrons, neutrons and isotopes, fill in the blanks below.

Symbol	Atomic Number	Isotope Mass	# of Protons	# of Electrons	# of Neutrons
	12	24	12	12	
C	6		6	6	7
Br		79	35	35	
	2		2	2	2
O	8	16			

Protons and neutrons make up an atom's nucleus. Below are simple diagrams of nuclei showing protons and neutrons.

Look up each element on the periodic table and then color the correct number of circles in the nucleus to represent how many protons there are. Count the circles left over--this is how many neutrons there are in the displayed isotope. Then color the neutrons.

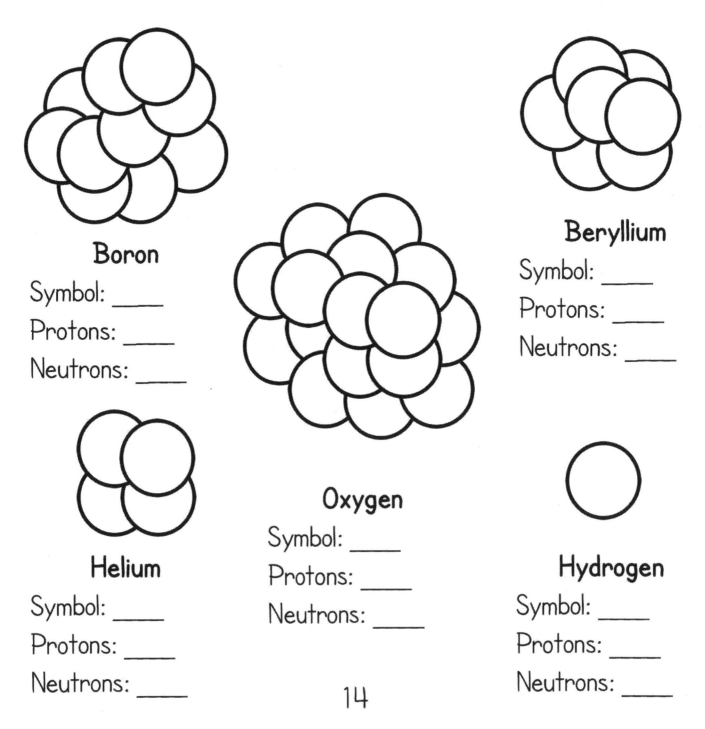

Boron

Symbol: _____

Protons: _____

Neutrons: _____

Beryllium

Symbol: _____

Protons: _____

Neutrons: _____

Helium

Symbol: _____

Protons: _____

Neutrons: _____

Oxygen

Symbol: _____

Protons: _____

Neutrons: _____

Hydrogen

Symbol: _____

Protons: _____

Neutrons: _____

Carbon

Symbol: ____

Protons: ____

Neutrons: ____

Magnesium

Symbol: ____

Protons: ____

Neutrons: ____

Chlorine

Symbol: ____

Protons: ____

Neutrons: ____

Aluminum

Symbol: ____

Protons: ____

Neutrons: ____

Here are more diagrams of nuclei, from the same elements as before, but this time of different isotopes. Color them like you did before, coloring first the protons and then counting the remaining circles to determine the number of neutrons in these isotopes.

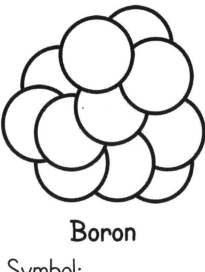

Boron

Symbol: ____

Protons: ____

Neutrons: ____

Beryllium

Symbol: ____

Protons: ____

Neutrons: ____

Oxygen

Symbol: ____

Protons: ____

Neutrons: ____

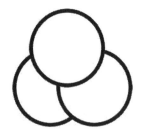

Helium

Symbol: ____

Protons: ____

Neutrons: ____

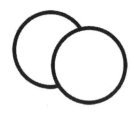

Hydrogen

Symbol: ____

Protons: ____

Neutrons: ____

Carbon

Symbol: ____

Protons: ____

Neutrons: ____

Magnesium

Symbol: ____

Protons: ____

Neutrons: ____

Chlorine

Symbol: ____

Protons: ____

Neutrons: ____

Aluminum

Symbol: ____

Protons: ____

Neutrons: ____

Orbits and Orbitals

A simple way of understanding the behavior of electrons is the Bohr model of the atom. Niels Bohr came up with a solar system model to introduce the concept that electrons only exist in certain orbits.

Below is the Bohr model of a calcium atom:

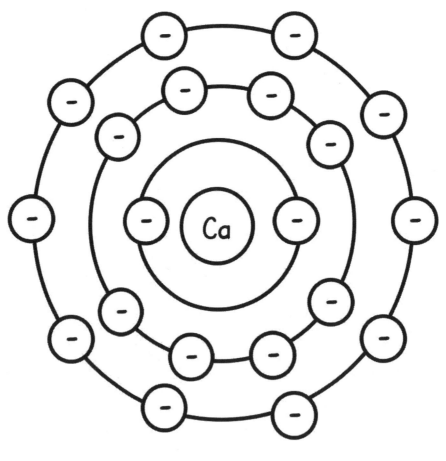

Calcium has:

___ protons

___ electrons

According to the Bohr model, each orbit around the nucleus can have a set number of electrons. Orbits closest to the nucleus are filled first. After each orbit is filled, additional electrons go to the next closest orbit.

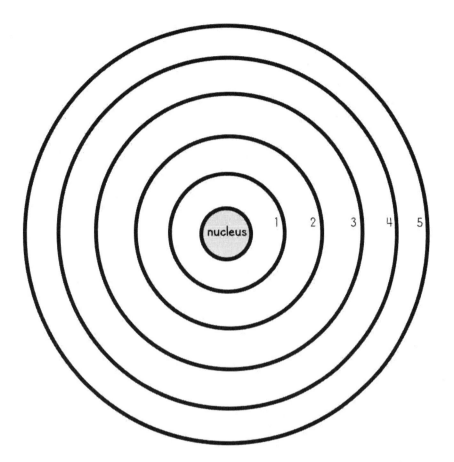

The way to calculate how many electrons an orbit can hold is $2n^2$ where n is the orbit number

1st orbit: $2(1^2)$ = 2 electrons
2nd orbit: $2(2^2)$ = 8 electrons
3rd orbit: $2(3^2)$ = 18 electrons
4th orbit: $2(4^2)$ = 32 electrons
5th orbit: $2(5^2)$ = 50 electrons

Draw electrons on the orbits above, adding the correct number of electrons to each orbit.

Color the center circle representing the nucleus. Then draw electrons on the orbits around the nucleus, assigning the correct number of electrons to each orbit according to the Bohr model.

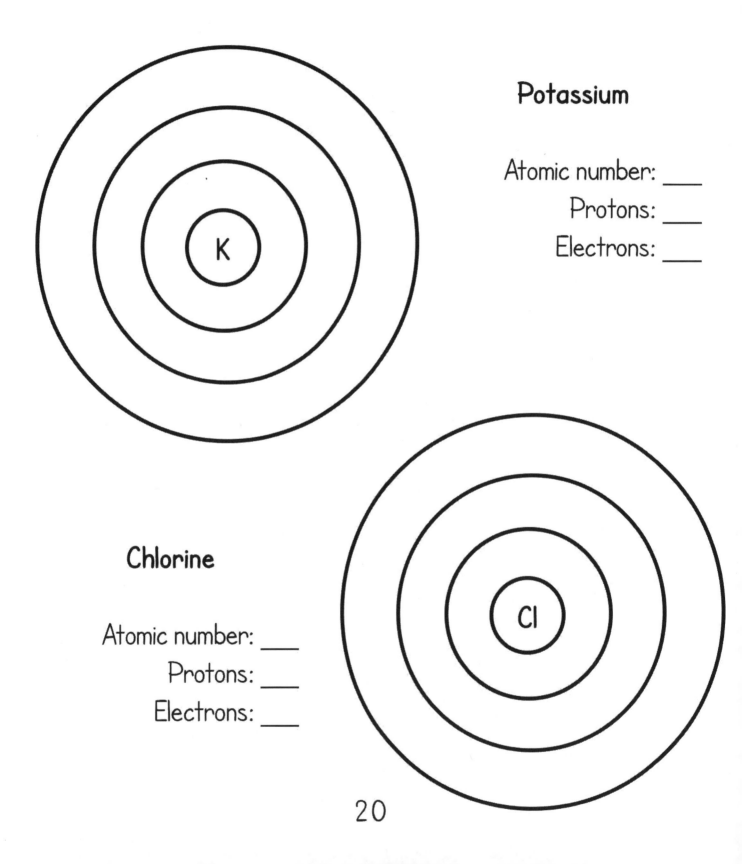

Potassium

Atomic number: ___
Protons: ___
Electrons: ___

Chlorine

Atomic number: ___
Protons: ___
Electrons: ___

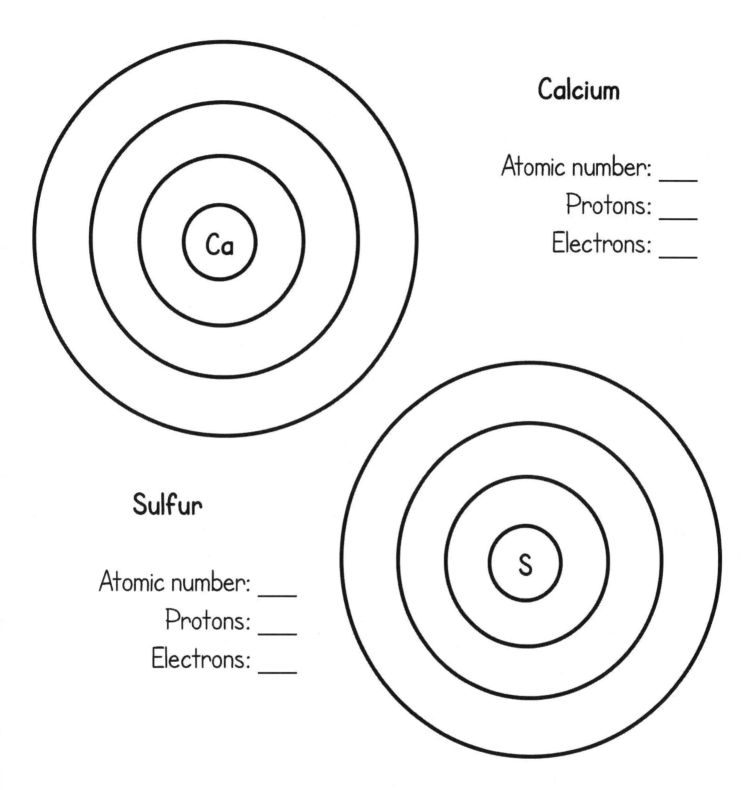

Calcium

Atomic number: ___
Protons: ___
Electrons: ___

Sulfur

Atomic number: ___
Protons: ___
Electrons: ___

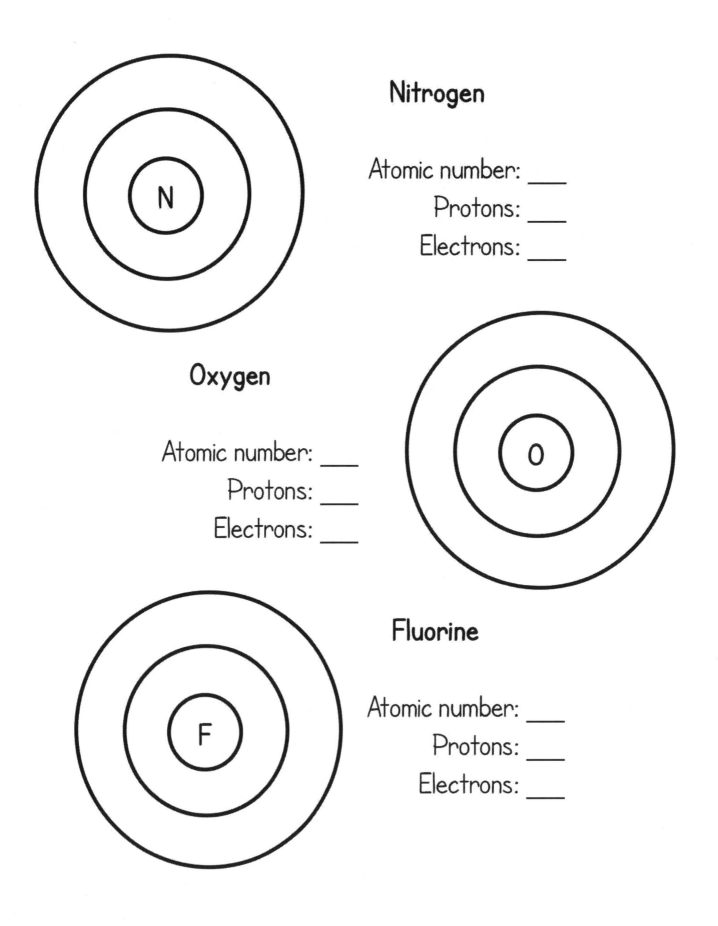

Nitrogen

Atomic number: ___

Protons: ___

Electrons: ___

Oxygen

Atomic number: ___

Protons: ___

Electrons: ___

Fluorine

Atomic number: ___

Protons: ___

Electrons: ___

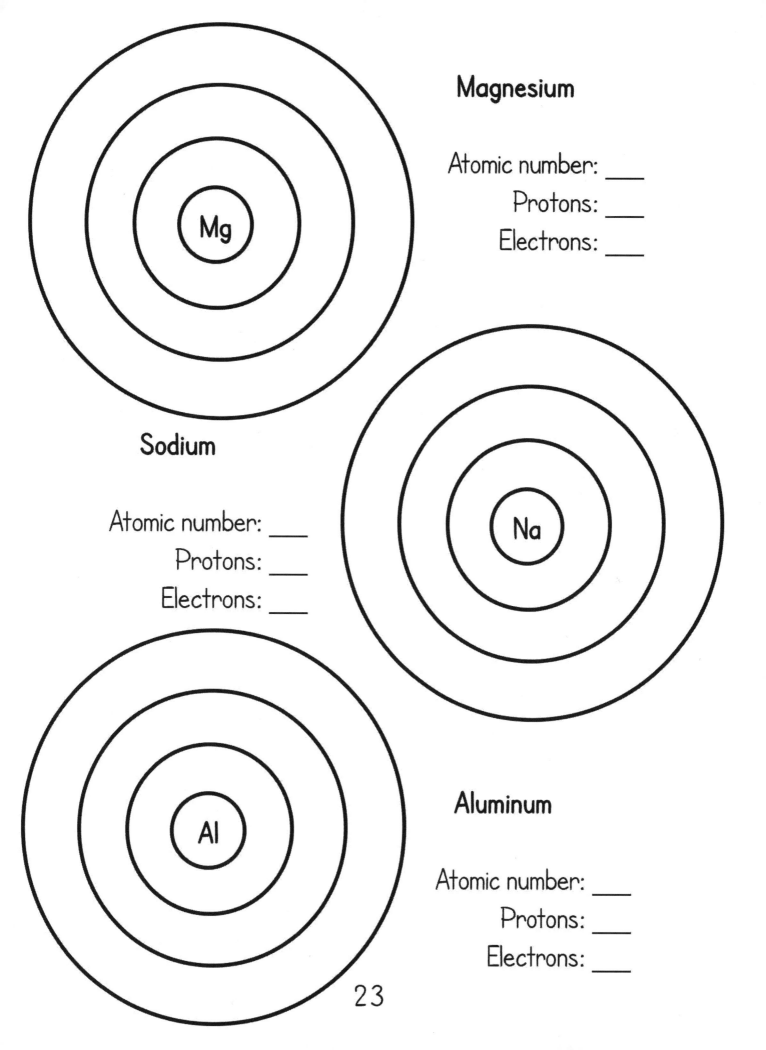

Magnesium

Atomic number: ___

Protons: ___

Electrons: ___

Sodium

Atomic number: ___

Protons: ___

Electrons: ___

Aluminum

Atomic number: ___

Protons: ___

Electrons: ___

23

This time pay attention to the charge when adding electrons to the models. The charge tells you how many electrons the atom has gained or lost.

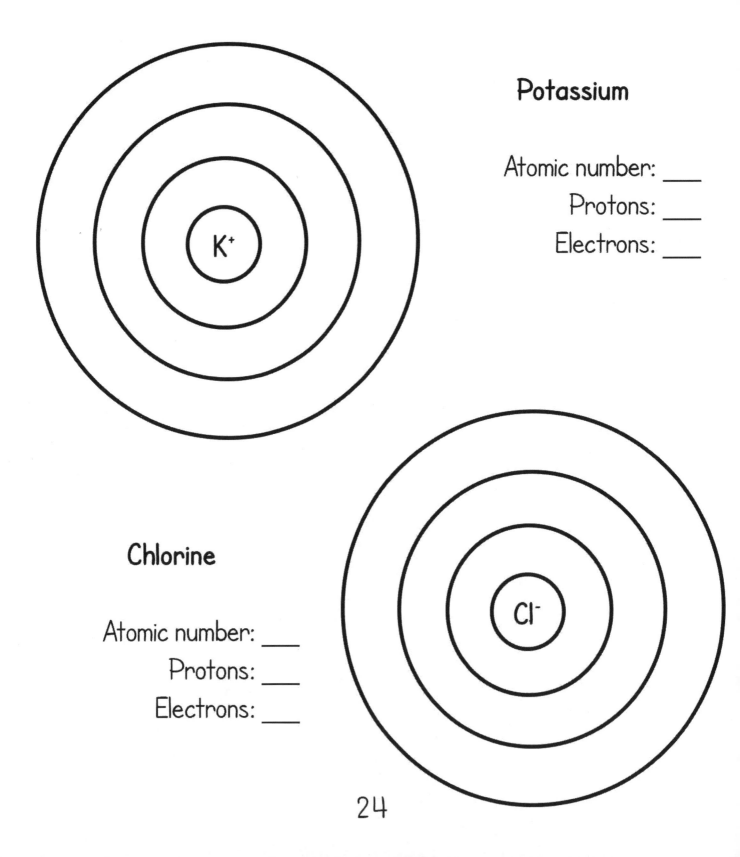

Potassium

Atomic number: ___
Protons: ___
Electrons: ___

Chlorine

Atomic number: ___
Protons: ___
Electrons: ___

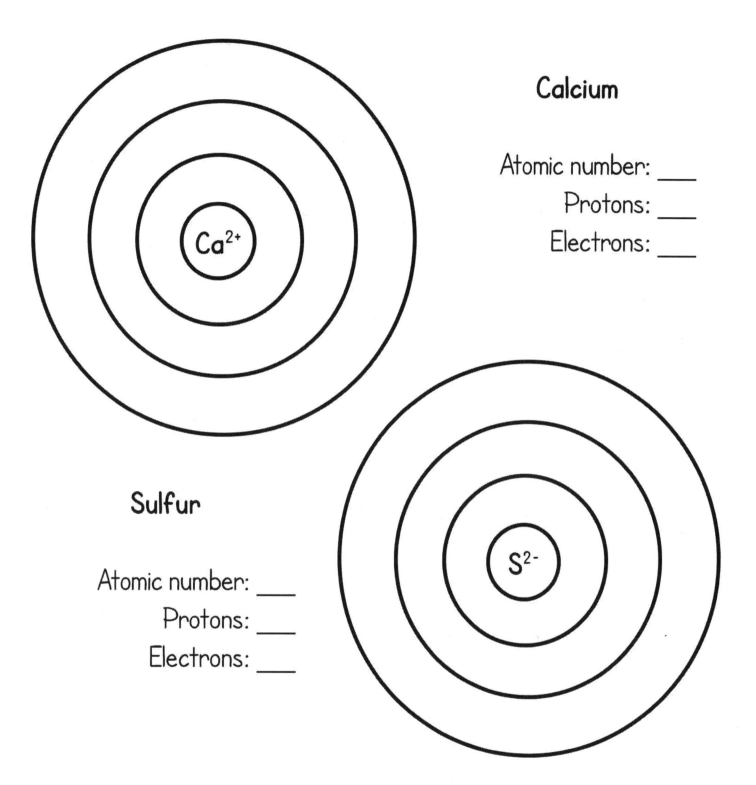

Calcium

Atomic number: ___
Protons: ___
Electrons: ___

Sulfur

Atomic number: ___
Protons: ___
Electrons: ___

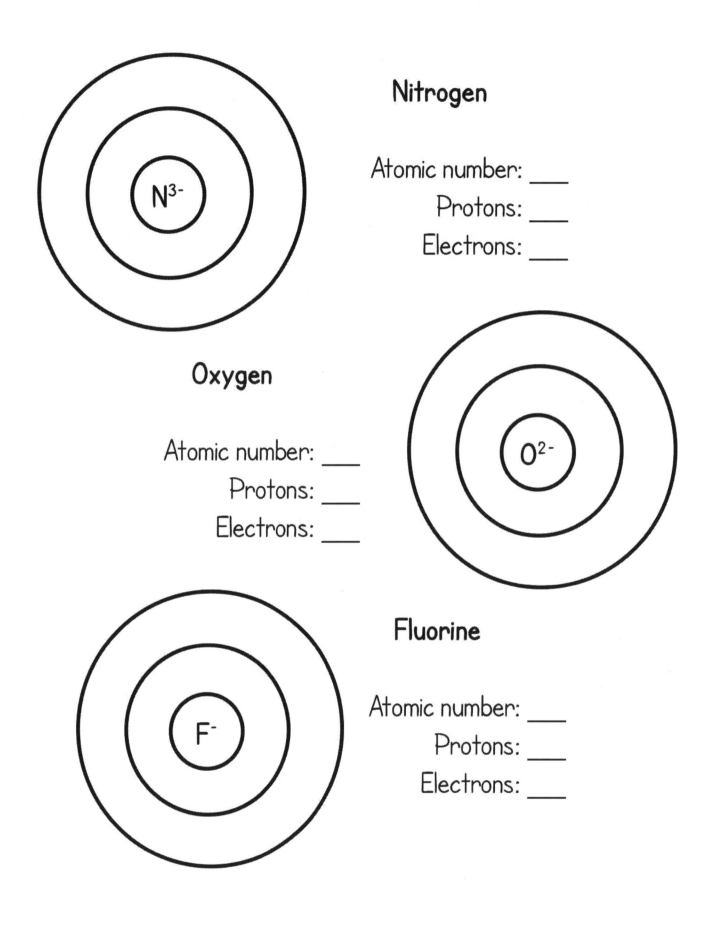

Nitrogen

Atomic number: ___
Protons: ___
Electrons: ___

Oxygen

Atomic number: ___
Protons: ___
Electrons: ___

Fluorine

Atomic number: ___
Protons: ___
Electrons: ___

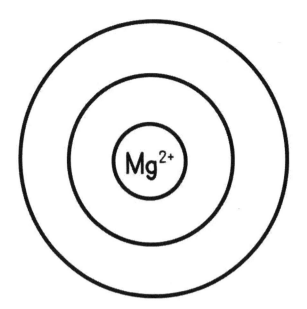

Magnesium

Atomic number: ___

Protons: ___

Electrons: ___

Sodium

Atomic number: ___

Protons: ___

Electrons: ___

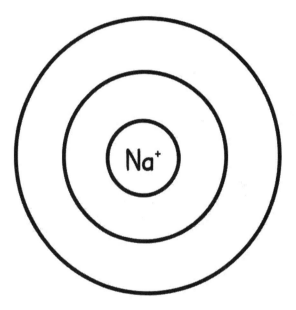

Aluminum

Atomic number: ___

Protons: ___

Electrons: ___

While the Bohr model is useful for diagramming an atom's electrons, it is not the most accurate model for showing the ways electrons behave in reality.

According to the **Heisenberg uncertainty principle**, either an electron's position or velocity may be measured at one time, but both can not be measured simultaneously.

Rather than moving in predictable, circular orbits, electrons exist in electron shells, called **orbitals**, of various shapes. Each orbital can hold 2 electrons.

There are four known orbital shapes: *s, p, d,* and *f*

The *s* orbitals have a spherical shape and can hold 2 electrons each.

The *p* orbitals have a dumbbell shape and there are 3 orbitals on each *p* sublevel, so each *p* sublevel can hold 6 electrons.

The *d* orbitals have a cloverleaf shape and there are five on each *d* sublevel. Each *d* sublevel can hold 10 electrons.

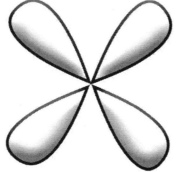

The *f* orbitals have a double cloverleaf shape and there are seven on each *f* sublevel. Each *f* sublevel can hold 14 electrons.

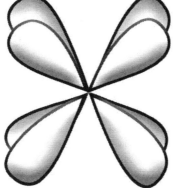

Choose a color for each type of orbital and color them in. Be sure to choose colors that will work well for coloring another copy of the periodic table.

Electrons fill orbitals with the lowest energy first. The sequence from lowest to highest energy is:

1s, 2s, 2p, 3s, 3p, 4s, 3d, 4p, 5s, 4d, 5p, 6s, 4f, 5d, 6p, 7s, 5f, 6d, 7p

The diagram below can help you remember the correct order in which the orbitals are filled.

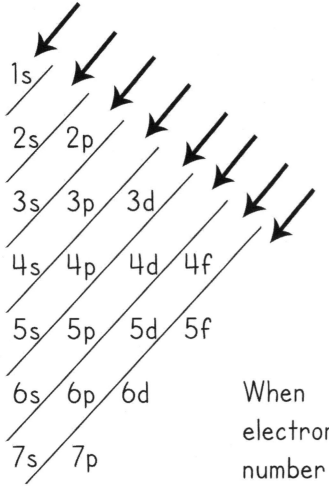

The 118th element on the periodic table would have the folowing configuration with all electron shells filled:

$1s^2$, $2s^2$, $2p^6$, $3s^2$, $3p^6$, $4s^2$, $3d^{10}$, $4p^6$, $5s^2$, $4d^{10}$, $5p^6$, $6s^2$, $4f^{14}$, $5d^{10}$, $6p^6$, $7s^2$, $5f^{14}$, $6d^{10}$, $7p^6$

When determining an element's electronic configuration, we use the number of electrons established by the element's atomic number

Hydrogen's electronic configuration therefore would be $1s^1$ and helium's is $1s^2$. The first s orbital is filled first. The third element on the periodic table, lithium, has its outermost electron in the 2s shell. Lithium's electronic configuration is $1s^2$, $2s^1$. As each shell is filled, the next electron moves to the next closest orbital.

Remember that s sublevels can hold 2 electrons each, p sublevels can hold 6 electrons each, d sublevels can hold 10 electrons each, and f sublevels can hold up to 14 electrons each.

Cobalt's atomic number is 27. See if you can fill in the blanks to find cobalt's electronic configuration.

Cobalt: $1s^2$, ___, $2p^6$, $3s^2$, ___, $4s^2$, $3d^7$

You should have written $2s^2$ and $3p^6$. Notice that the last number shown is $3d^7$, not $3d^{10}$. This is because cobalt only has 7 electrons in its outer shell, even though d orbitals can hold up to 10 electrons. Look at the periodic table and see if you can figure out which element has an outer shell of $3d^{10}$.

Try filling in a few more electronic configurations:

Oxygen: ___, $2s^2$, $2p^4$

Strontium: $1s^2$, $2s^2$, ___, ___, $3p^6$, $4s^2$, ___, $4p^6$, ___

Phosphorus: $1s^2$, ___, $2p^6$, $3s^2$, ___

Here are the correct configurations:

Oxygen: $1s^2$, $2s^2$, $2p^4$
Strontium: $1s^2$, $2s^2$, $2p^6$, $3s^2$, $3p^6$, $4s^2$, $3d^{10}$, $4p^6$, $5s^2$
Phosphorus: $1s^2$, $2s^2$, $2p^6$, $3s^2$, $3p^3$

This periodic table shows what type of orbital each element's most outer electron occupies. Color the periodic table using the same colors you used for the orbital diagrams on pages 28-29. Use those same colors for orbitals throughout the book.

Orbital Orientation

It was mentioned before that there are 3 p orbitals on each p sublevel, 5 d orbitals on each d sublevel, and 7 f orbitals on each f sublevel. These orbitals are aligned to axes within their respective sublevels.

The p orbitals are aligned on the x, y and z axes.

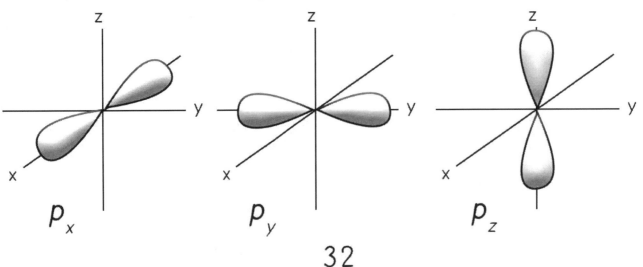

P_x P_y P_z

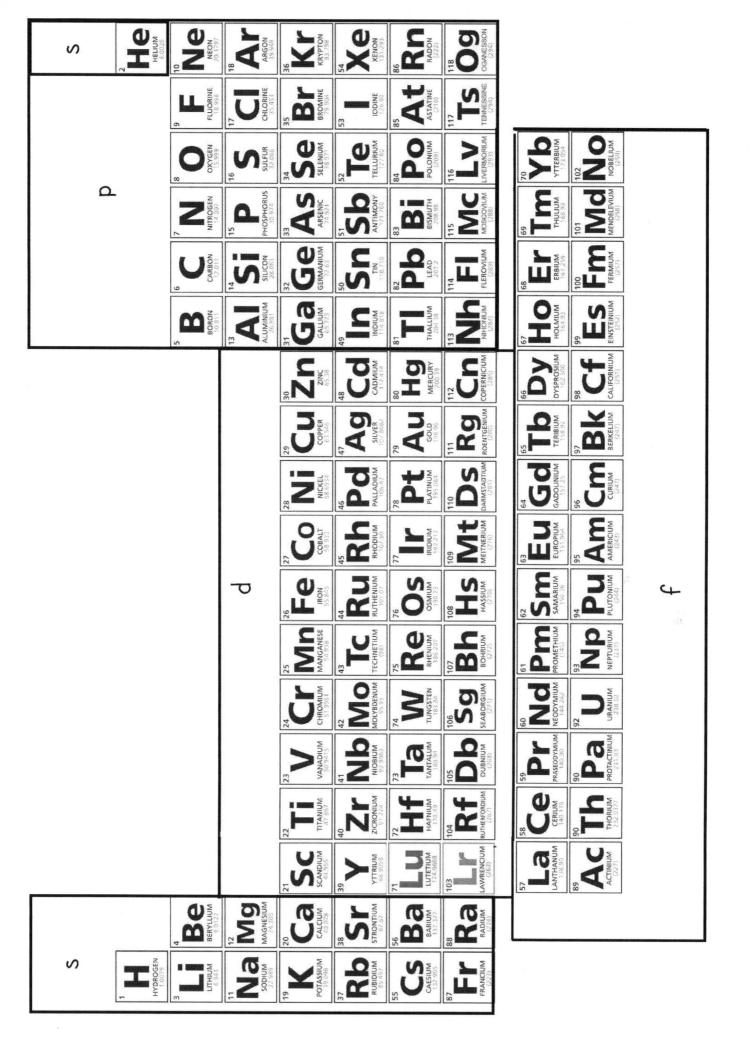

Here are the shapes of the *d* orbitals. Four of them are identical cloverleaf shapes; the fifth is a different shape.

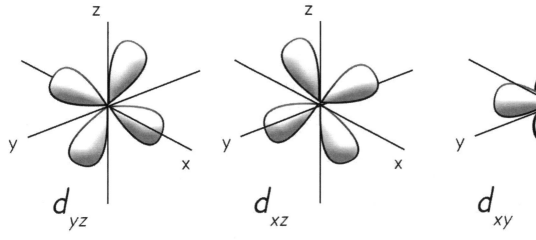

d_{yz}

d_{xz}

d_{xy}

$d_{x^2-y^2}$

d_{z^2}

Here are the shapes of the *f* orbitals.

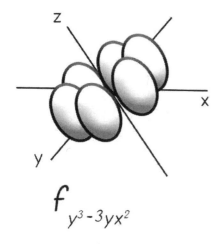

$f_{y^3-3yx^2}$

f_{xyz}

$f_{5yz^2-yr^2}$

34

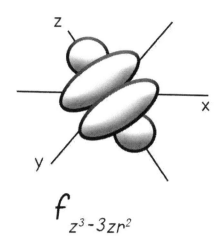

$f_{z^3-3zr^2}$

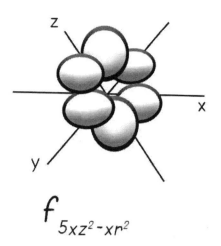

$f_{5xz^2-xr^2}$

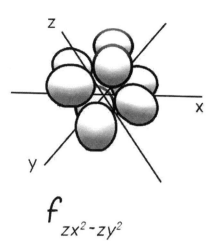

$f_{zx^2-zy^2}$

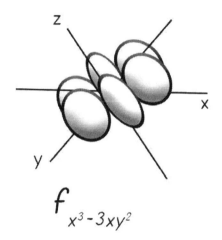

$f_{x^3-3xy^2}$

To review:

Each orbital can hold up to 2 electrons. There is one orbital in each s sublevel. There are three orbitals in each p sublevel, five orbitals in each d sublevel, and seven orbitals in each f sublevel.

Valence Electrons

Another word for the outermost electrons in an atom, particularly in the *s* and *p* orbitals, is **valence** electrons. Valence electrons in the *s* and *p* orbitals play an important role in chemical bonding.

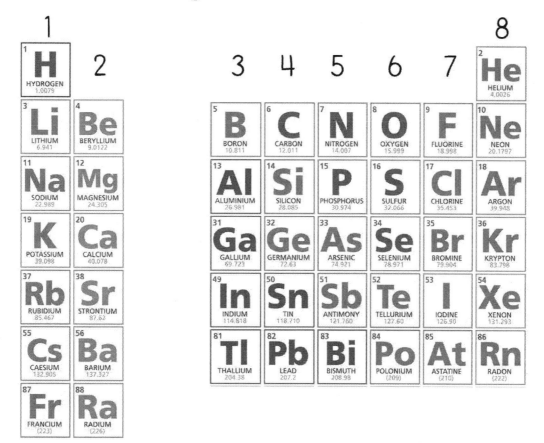

The diagram above shows how many valence electrons these columns of the periodic table have, with helium as an exception since helium has only 2 electrons in its outer shell. Color each column a different color

When atoms do not have a complete outer shell, they seek to either gain or lose electrons in order to reach this state.

The elements in the 8th column already have a complete outer shell. These elements are nonreactive and are called the **noble gases**.

Dots surrounding an element's atomic symbol can be used to represent valence electrons. Here are diagrams showing the valence electrons of one element from each column.

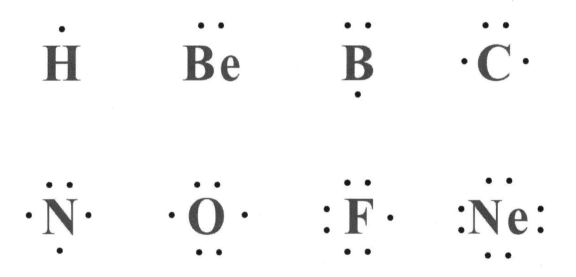

The positioning of the dots above is important. The two dots above beryllium are paired to represent the completed s orbital. This is why they are paired instead of balanced on opposite sides of the symbol.

After filling the *s* orbital, electrons begin to fill the *p* orbitals. According to **Hund's rule**, *p*, *d*, and *f* orbitals in a sublevel have to each be filled with a single electron before a second electron can pair in any orbital. This is why you don't see another pair of dots in the diagram until you get to oxygen.

Diagram the valence electrons for the elements below.

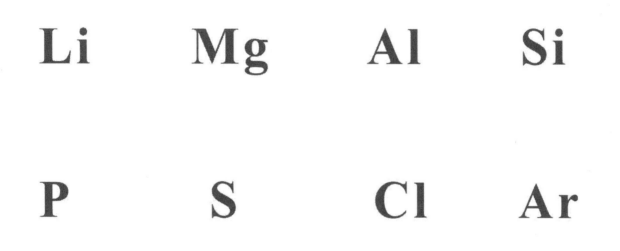

Another way to demonstrate Hund's rule is with orbital boxes. Orbital boxes show an empty orbital, an unpaired electron, or a pair of electrons, using arrows to represent the electronic spin.

Here are the first 10 elements of the periodic table shown using orbital box diagrams.

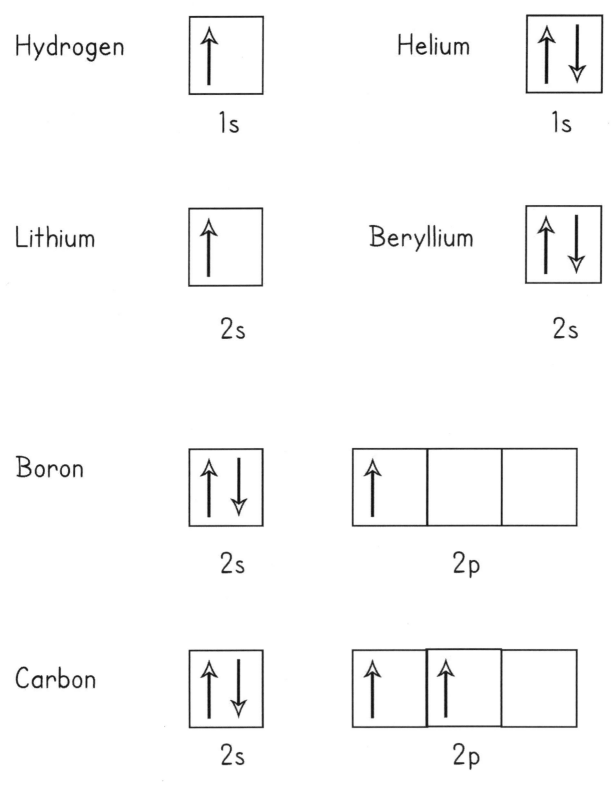

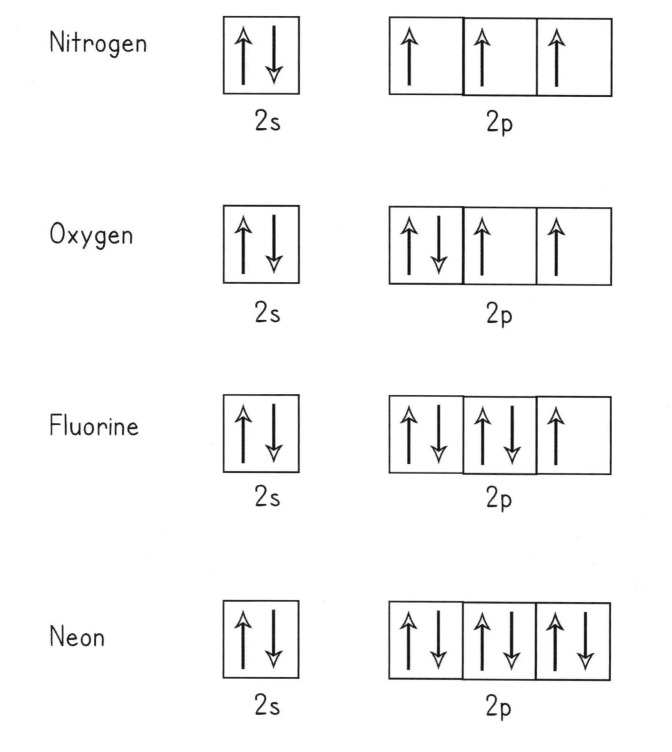

Here is the order of orbitals through the *d* sublevel:

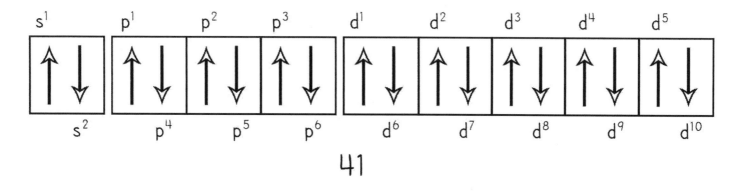

Complete the orbital boxes. See if you can remember where the arrows go without looking back at the answers.

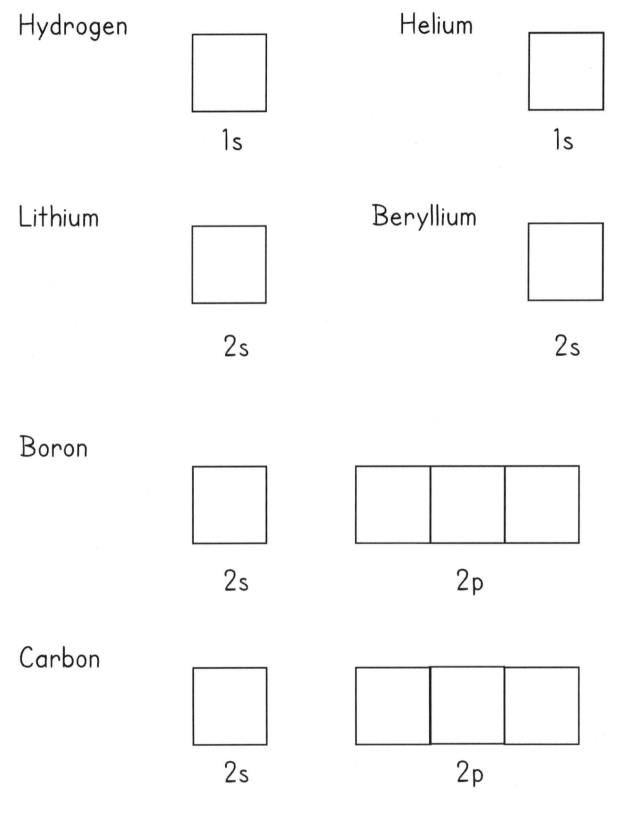

Hydrogen

1s

Helium

1s

Lithium

2s

Beryllium

2s

Boron

2s   2p

Carbon

2s   2p

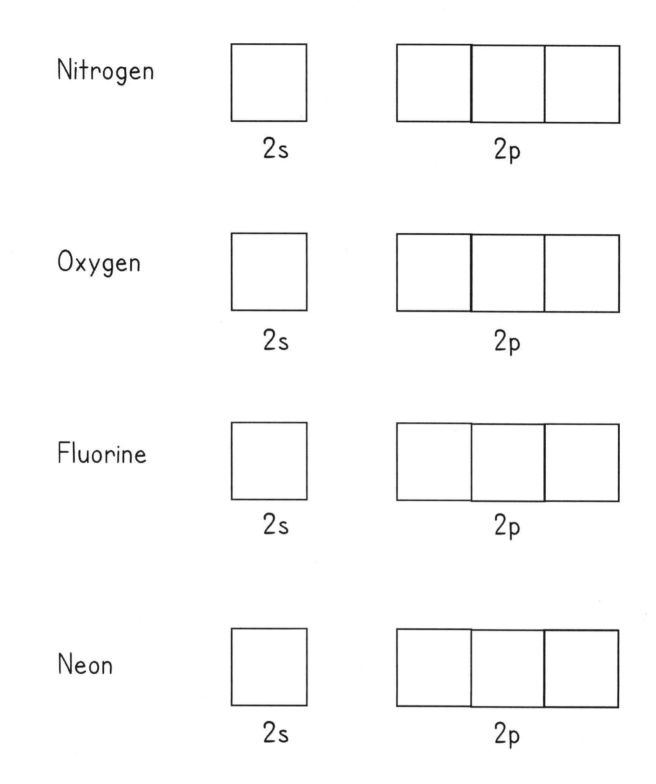

Nitrogen

2s   2p

Oxygen

2s   2p

Fluorine

2s   2p

Neon

2s   2p

Combining this information with the orbital shapes, color the orbital diagrams on pages 44-45 and fill in the blanks below each to show which elements have their outermost electron in that orbital. The first 18 are done for you.

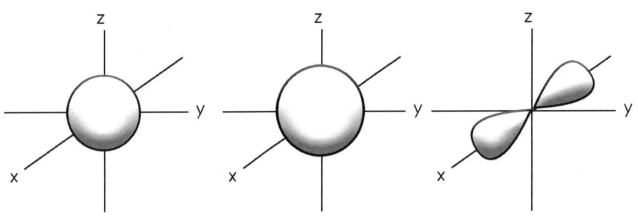

$1s^1$ $1s^2$ $2s^1$ $2s^2$ $2p^1$ $2p^4$
H   He   Li   Be   B   O

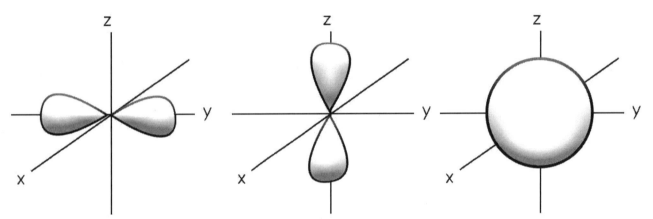

$2p^2$ $2p^5$ $2p^3$ $2p^6$ $3s^1$ $3s^2$
C   F   N   Ne   Na   Mg

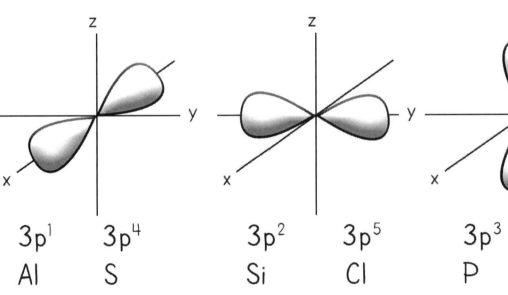

$3p^1$ $3p^4$ $3p^2$ $3p^5$ $3p^3$ $3p^6$
Al   S   Si   Cl   P   Ar

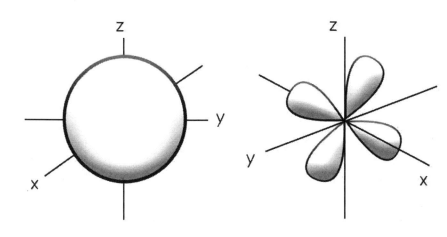

$4s^1$ $4s^2$ $3d^1$ $3d^6$ $3d^2$ $3d^7$

––   ––   ––   ––   ––   ––

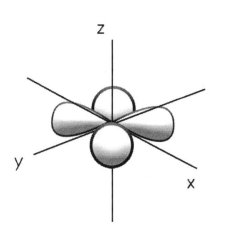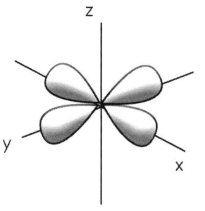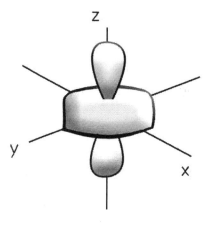

$3d^3$ $3d^8$ $3d^4$ $3d^9$ $3d^5$ $3d^{10}$

– –   ––   ––   ––   ––   ––

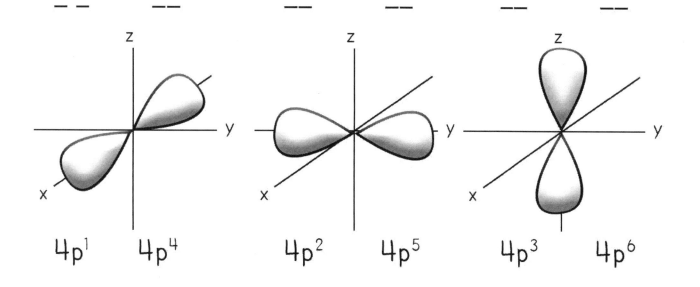

$4p^1$ $4p^4$ $4p^2$ $4p^5$ $4p^3$ $4p^6$

––   ––   ––   ––   ––   ––

The remainder of this book explores chemical bonding. To best describe chemical bonds, more terms will be used to clarify sections of the periodic table. You have already colored the periodic table divided into metals, nonmetals, and metalloids. Now we will take those divisions a step further

Color the key using a different color for each box.

Color the following with the color you chose for nonmetals:
H, C, N, O, P, S, Se

The alkali metals are the metals below hydrogen in the first column:
Li, Na, K, Rb, Cs, Fr

The alkaline earth metals are in the second column:
Be, Mg, Ca, Sr, Ba, Ra

The metalloids are the same as you colored before:
B, Si, Ge, As, Sb, Te, Po

The noble gases are in the column on the far right:
He, Ne, Ar, Kr, Xe, Rn, Og

The halogens are in the column beside the noble gases:
F, Cl, Br, I, At, Ts

The Lanthanides are numbers 57-71
The Actinides are numbers 89-103

Color the following with the color you chose for metals:
Al, Ga, In, Sn, Tl, Pb, Bi, Nh, Fl, Mc, Lv

The remaining elements, which all have their outer electrons in *d* orbitals, are transition metals

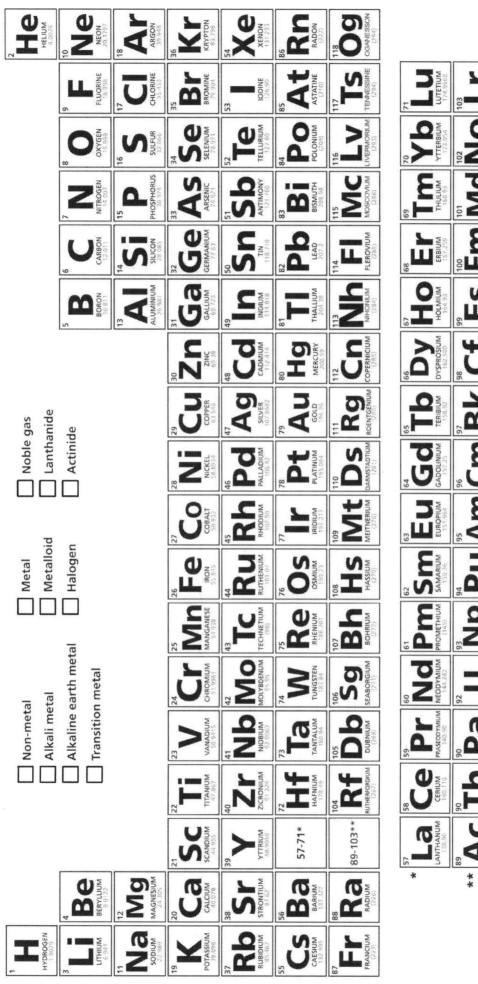

Chemical Bonding

When nonmetals bond together they **share** electrons, forming **covalent** bonds. All of the nonmetals except for hydrogen and helium have their valence electrons in p orbitals. They like to gain electrons to fill their outer p sublevel, except for the noble gases which already have their orbitals full.

Metals, on the other hand, bond with nonmetals by giving up electrons rather than sharing electrons. These are called **ionic** bonds. The alkali metals and the alkaline earth metals are the most reactive because their valence electrons are in s orbitals. These elements like to give up their s electrons completely, leaving them with a completed outer shell in their outermost p sublevel. Exceptions are lithium and beryllium, which do not have any electrons in p orbitals, Their outer electrons are in the 2s sublevel; they give up these electrons, leaving them with their outermost electrons in the 1s sublevel.

Hydrogen, being a nonmetal in the first column of the periodic table, sometimes forms covalent bonds with other nonmetals and at other times behaves more like a metal and forms ionic bonds.

In the following pages you'll learn about and color representations of chemical compounds. Color the table below as a reference to look back at when coloring the chemical structures. Use these colors to represent these elements for the rest of the book.

Hydrogen - white
Carbon - black
Oxygen - red
Nitrogen - light blue
Fluorine - orange
Bromine - dark blue

Chlorine - green
Phosphorus - purple
Sulfur - yellow
Iodine - lavender
Metals - gray

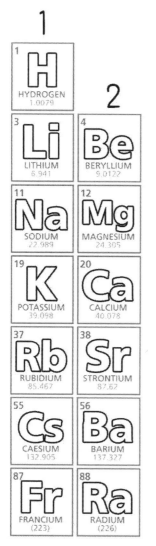

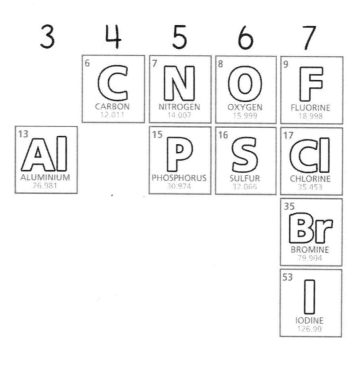

Diatomic Elements

There are seven elements which are never found as a single solitary atom. They are always paired up with another atom to form a molecule, and in a solution of only that element, they will be paired with each other.

The diatomic elements are H, N, O, F, Cl, Br, I

Notice that four of these are halogens. The halogens each have 7 electrons in their valence shell. They want to have a complete outer shell, so they seek to gain an electron to fill it. Because of this, atoms of these elements form covalent bonds to share an electron with each other

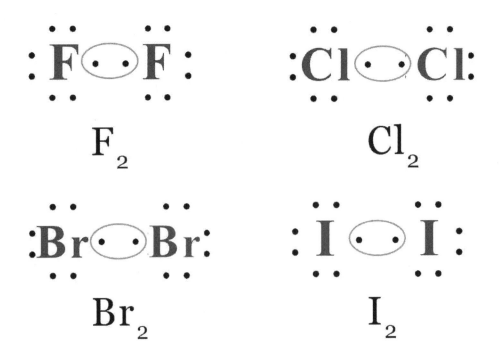

Hydrogen shares its only electron to form hydrogen gas.

H$_2$

Oxygen has 6 electrons in its valence shell. It shares 2 electrons to complete its outer shell. This is a double bond

O$_2$

Nitrogen has only 5 electrons in its valence shell. Nitrogen shares 3 electrons to complete its outer shell. This is a triple bond.

:N$\;$::$\;$:N: N$_2$

Color the diatomic molecules.

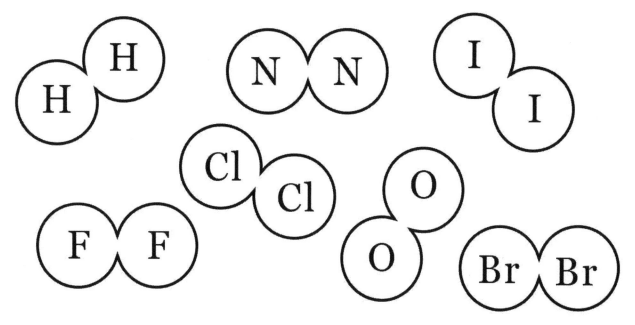

Two molecules of hydrogen gas can react with a molecule of oxygen gas to form two molecules of water

$$2H_2 + O_2 = 2H_2O$$

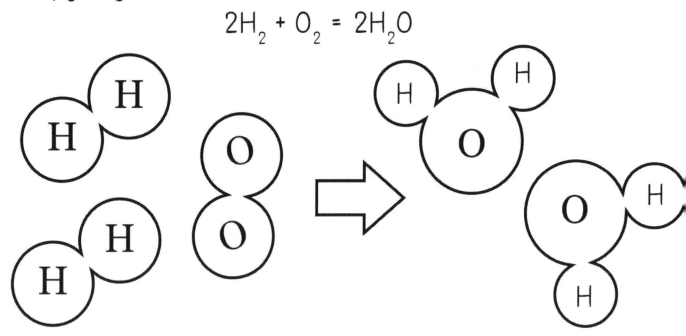

One molecule of hydrogen gas can react with one molecule of oxygen gas to form a molecule of hydrogen peroxide.

$$H_2 + O_2 = H_2O_2$$

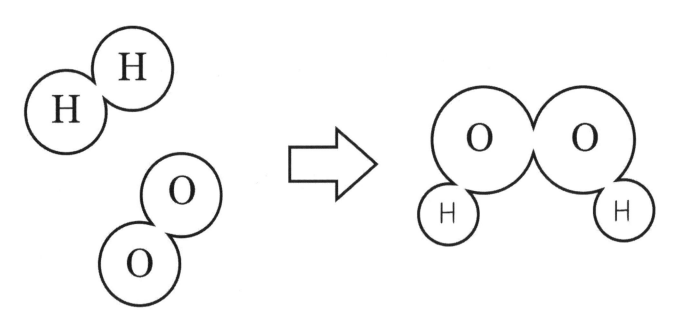

Ionic Compounds

Ionic bonds form between metals and nonmetals, and sometimes between hydrogen and other nonmetals.

In ionic bonds, electrons are transferred from one atom or compound to another, instead of being shared among atoms like with covalent bonds.

The metals we'll focus on in this book are the alkali metals, the alkaline earth metals, and aluminum.

The alkali metals each have a single valence electron in their outermost s orbital. They like to give away this electron to have a full outer shell. The alkali metals in their ionic form have a charge of 1+

The alkaline earth metals each have two valence electrons in their outermost s orbital. They like to give away these two electrons to have a full outer shell. When ions, the alkaline earth metal ions have a charge of 2+

Aluminum has three electrons in its outermost p orbital. It likes to give away these electrons. Aluminum ions have a charge of 3+

Alkali Metals + Halogens

The alkali metals in their ionic form have a charge of 1+ and the halogens in their ionic form have a charge of 1- One halogen atom and one alkali metal atom form an ionic bond together to create a balanced compound.

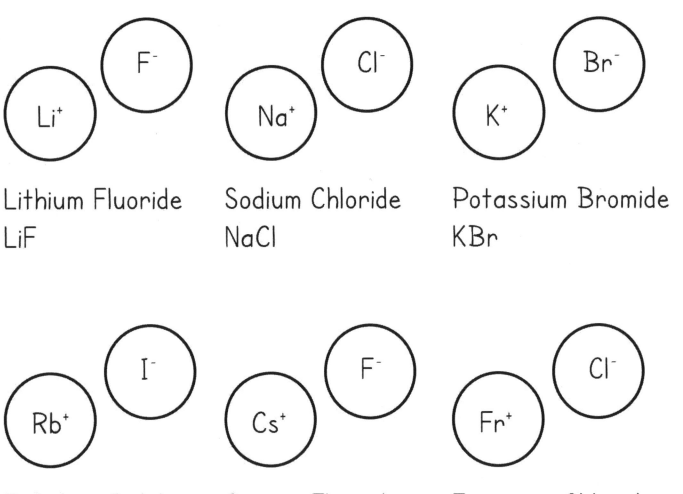

Lithium Fluoride
LiF

Sodium Chloride
NaCl

Potassium Bromide
KBr

Rubidium Iodide
RbI

Cesium Fluoride
CsF

Francium Chloride
FrCl

Alkali Metals + Oxygen

An oxygen ion has a charge of 2-. It takes two atoms of an alkali metal to bond with oxygen.

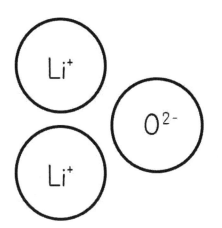

Lithium Oxide
Li$_2$O

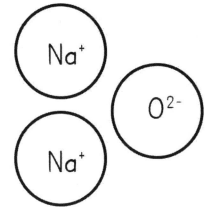

Sodium Oxide
Na$_2$O

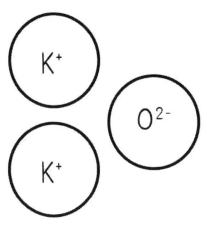

Potassium Oxide
K$_2$O

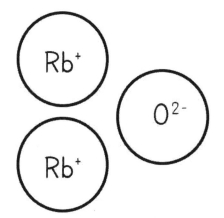

Rubidium Oxide
Rb$_2$O

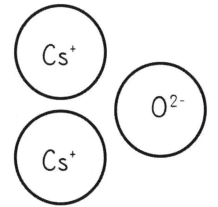

Cesium Oxide
Cs$_2$O

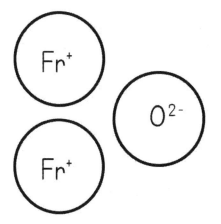

Francium Oxide
Fr$_2$O

Alkaline Earth Metals + Halogens

The alkaline earth metals in their ionic form have a charge of 2-. Halogens have a charge of 1-. It takes two atoms of a halogen to bond with an alkaline earth metal.

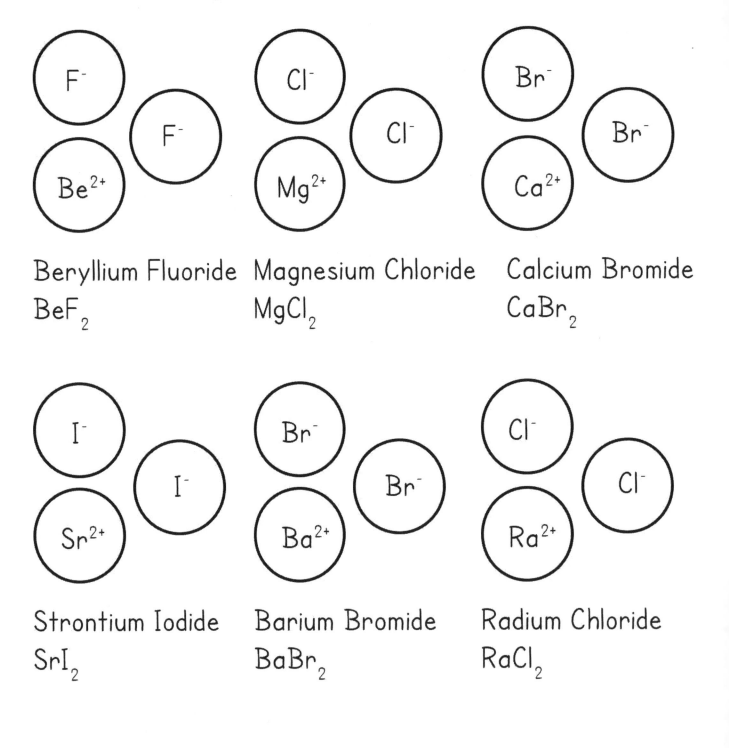

Beryllium Fluoride
BeF_2

Magnesium Chloride
$MgCl_2$

Calcium Bromide
$CaBr_2$

Strontium Iodide
SrI_2

Barium Bromide
$BaBr_2$

Radium Chloride
$RaCl_2$

Alkaline Earth Metals + Oxygen

The alkaline earth metals in their ionic form have a charge of 2+ and oxygen in its ionic form has a charge of 2-. One alkaline earth metal atom and one oxygen atom can form an ionic bond together to create a balanced compound.

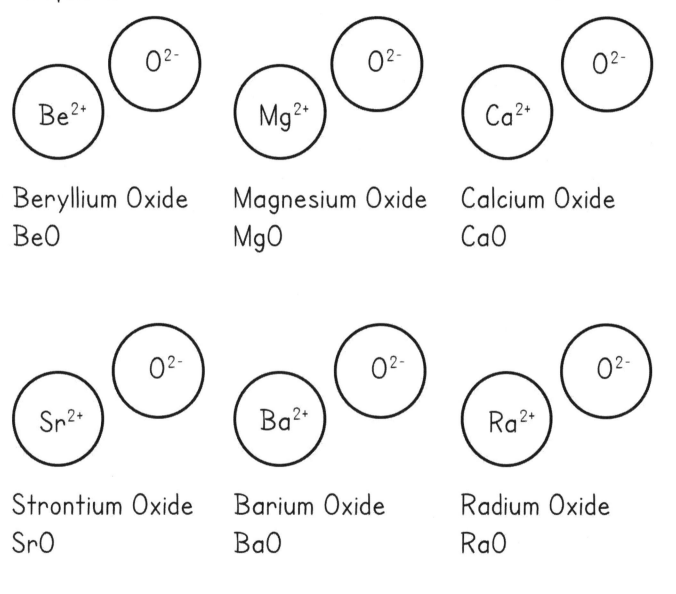

Beryllium Oxide
BeO

Magnesium Oxide
MgO

Calcium Oxide
CaO

Strontium Oxide
SrO

Barium Oxide
BaO

Radium Oxide
RaO

Aluminum + Halogens

Aluminum ions have a charge of 3+. Halogens have a charge of 1-. It takes three atoms of a halogen to bond with an aluminum atom.

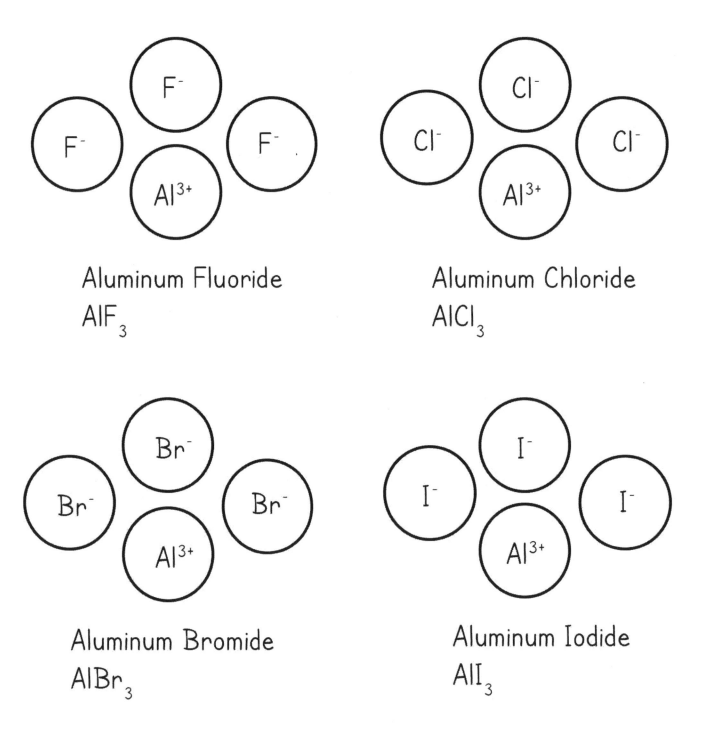

Aluminum Fluoride
AlF_3

Aluminum Chloride
$AlCl_3$

Aluminum Bromide
$AlBr_3$

Aluminum Iodide
AlI_3

Aluminum + Oxygen

Since aluminum has a charge of 3+ and oxygen has a charge of 2- it takes two aluminum atoms and three oxygen atoms to form a compound.

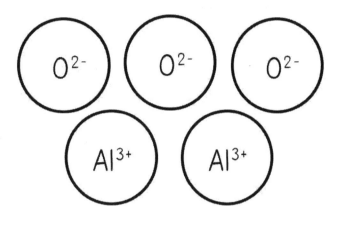

Aluminum Oxide

Al_2O_3

Complete these compounds by determining how many of the atoms should be the metal listed and how many should be of the other listed element.

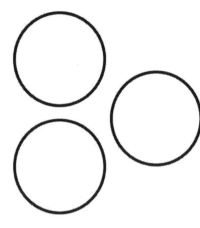

Potassium Oxide

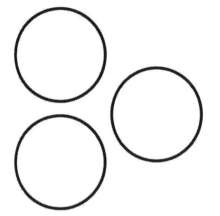

Barium Chloride

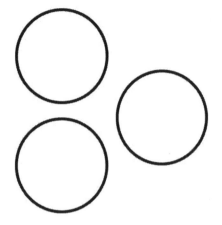

Calcium Fluoride

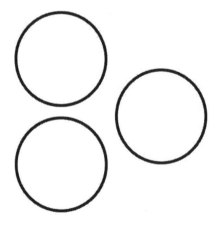

Radium Iodide

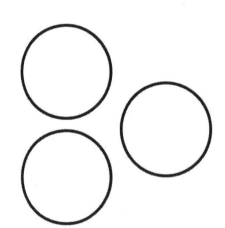

Sodium Oxide

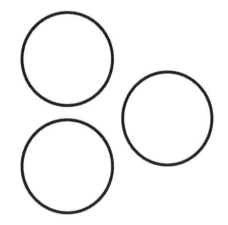

Strontium Bromide

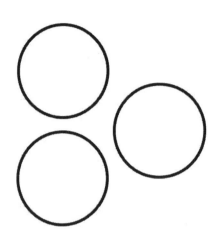

Magnesium Fluoride

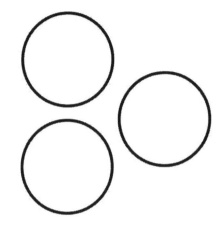

Calcium Iodide

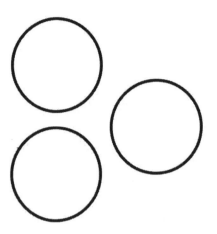

Lithium Oxide

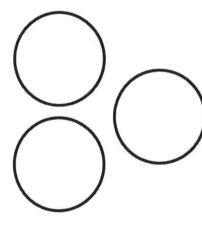

Cesium Oxide

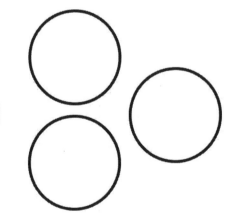

Beryllium Bromide

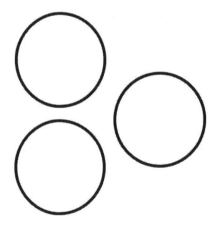

Magnesium Chloride

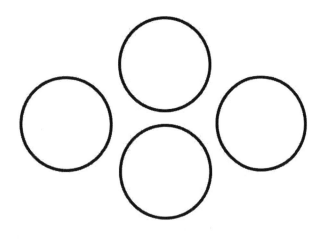

Aluminum Chloride

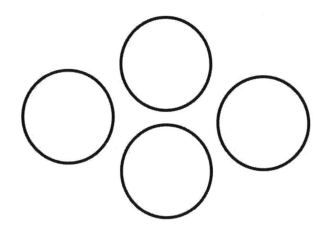

Aluminum Iodide

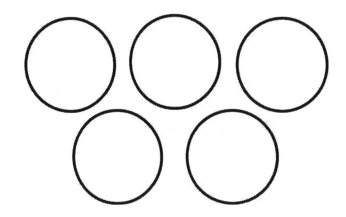

Aluminum Oxide

Covalent Compounds

In addition to the diatomic molecules, water, and hydrogen peroxide, here are some additional covalent molecules that can form. Notice that, like hydrogen and oxygen can bond in more than one way, so can the other elements form multiple variations of bonds with each other

Because covalent bonds involve sharing electrons, the bonds between the atoms are drawn. Compounds are shown in this book using several styles of ball-and-stick model. Atoms and bonds are not drawn to scale.

Beside each compound you will find either the lewis structure or the skeletal formula of that compound. These tell you which element each atom is so you can color correctly.

In skeletal formulas, carbon atoms are not always labeled. Often it will be shown as a corner where two bonds meet. Carbon is assumed to have 4 valence electrons. Any valence electrons not accounted for in the bonds shown can be assumed to be bonded to hydrogen atoms.

For the first four compounds both the lewis structure and the skeletal formula are shown so you can see the difference. After that, skeletal formulas are primarily shown.

Methane
CH$_4$

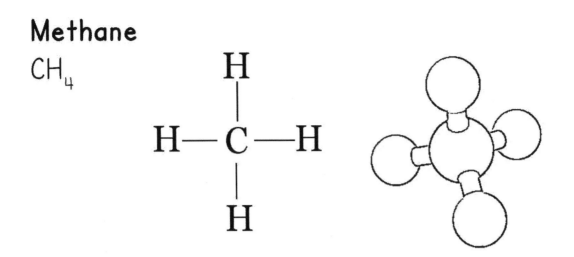

Carbon has 4 valence electrons. It likes to make 4 bonds to complete its valence shell. Carbon bonds with 4 hydrogen atoms to create methane.

Ethane
C$_2$H$_6$

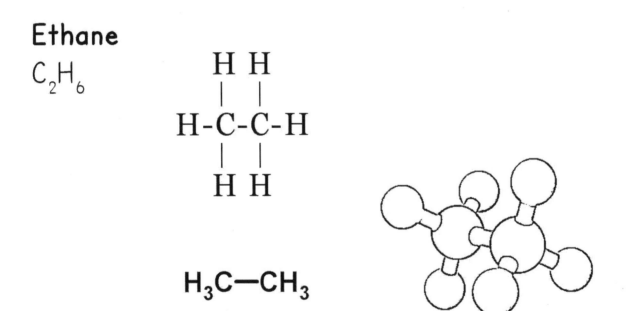

H$_3$C—CH$_3$

Two carbon atoms bond with 6 hydrogen atoms to create ethane. Ethane is made of two methyl groups. A methyl group is a carbon bonded to 3 hydrogen atoms.

Propane

C_3H_8

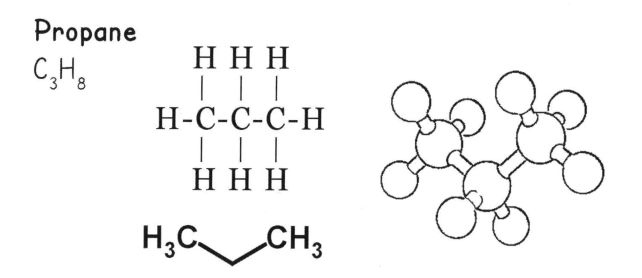

Propane has two methyl groups connected in the center by a methylene group. A methylene group is a carbon bonded to two hydrogen atoms. It has the formula CH_2

Butane

C_4H_{10}

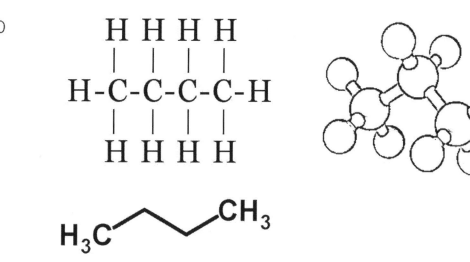

Butane has two methyl groups connected by two methylene groups.

Methanol

CH$_4$O

H$_3$C—OH

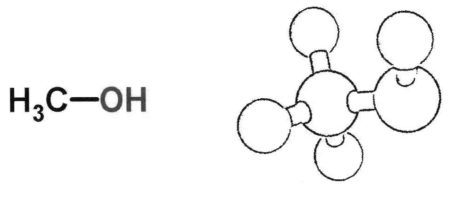

Methanol has a methyl group bonded to a hydroxyl group. A hydroxyl group is an oxygen atom bonded to a hydrogen atom. It has the formula OH. Oxygen has 6 valence electrons. It likes to bond to 2 electrons to complete its outer shell. In this example, oxygen has one bond to a carbon and one bond with a hydrogen.

Ethanol

C$_2$H$_6$O

H$_3$C〜OH

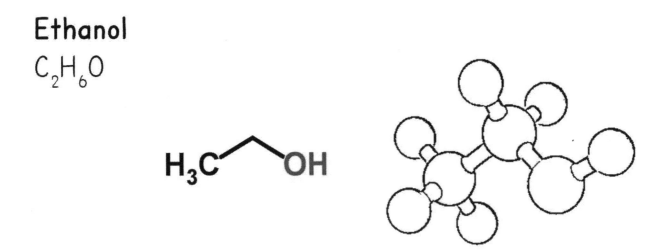

Ethanol has a methyl group bonded to a methylene group bonded to a hydroxyl group.

Propanol

C_3H_8O

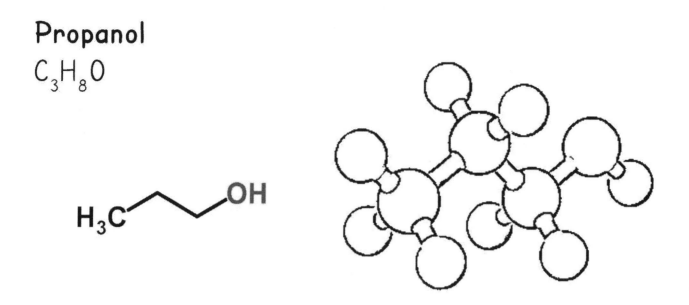

Propanol has a methyl group bonded to two methylene groups bonded to a hydroxyl group.

Butanol

$C_4H_{10}O$

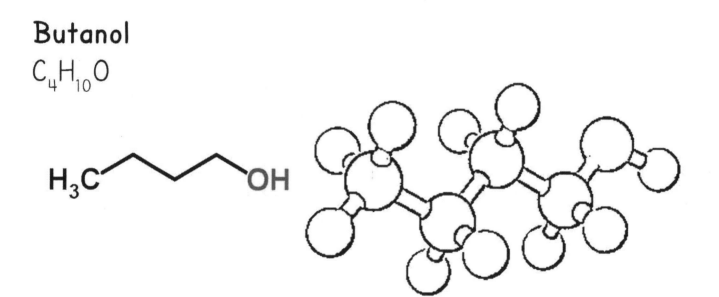

Butanol has a methyl group bonded to three methylene groups bonded to a hydroxyl group.

Try to draw the skeletal formulas for each of the following without looking back at the answers.

Methane

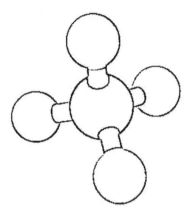

Ethane

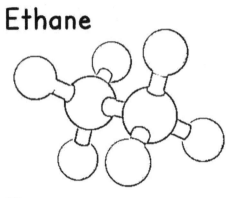

Propane

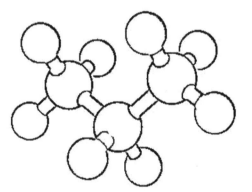

Butane

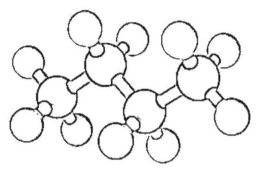

Methanol

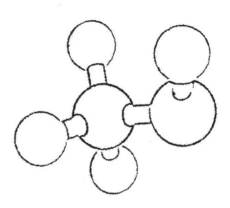

Ethanol

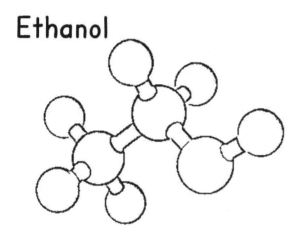

Propanol

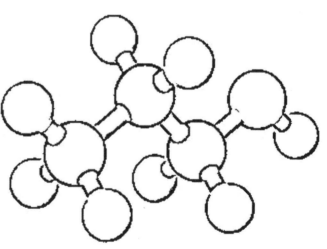

Butanol

Color the ball-and-stick models of each functional group.
Then fill in the blanks with the name of each.

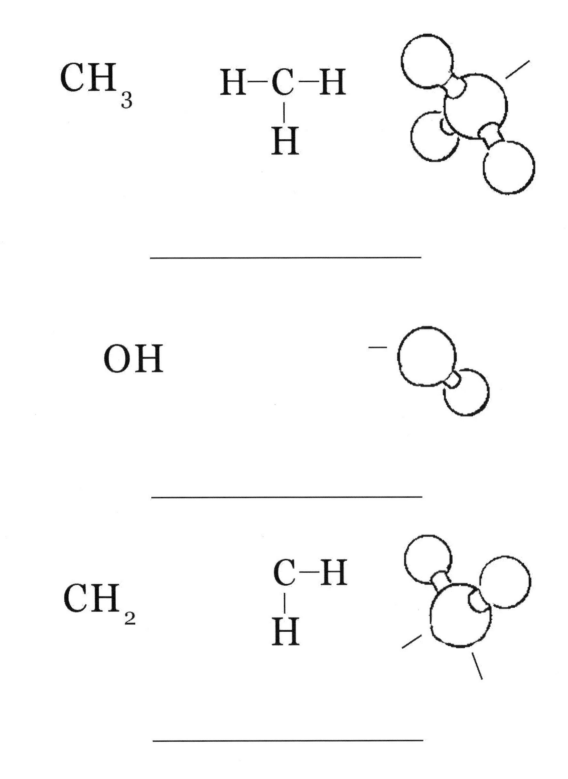

CH₃   H–C–H
            |
            H

_____

OH

_____

CH₂   C–H
         |
         H

_____

Carbon dioxide
CO_2

$$O=C=O$$

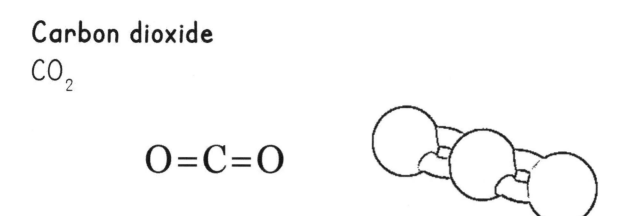

Carbon shares its 4 electrons with 2 oxygen atoms which each want a double bond. In skeletal formulas, a double bond is indicated by two lines.

Oxygen gas
O_2

$$O=O$$

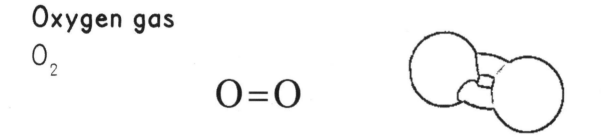

Oxygen gas, as mentioned previously, is a diatomic element. When oxygen gas forms, the oxygen atoms also have a double bond with each other.

When we breathe, our body uses oxygen gas and we exhale carbon dioxide. Plants use the carbon dioxide we exhale and they release oxygen gas into the air

Ammonia

NH_3

H H
 \ /
  N
  |
  H

Nitrogen has 5 valence electrons. It likes to create covalent bonds where 3 electrons are shared. This gives nitrogen a complete outer shell. Nitrogen bonds with 3 hydrogen atoms to form ammonia. Hydrogen only wants one electron to complete its outer shell. It shares its one electron and an electron is shared with it in return.

Nitrogen gas

N_2

$$N \equiv N$$

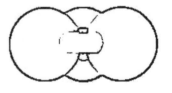

When paired with itself as a diatomic element, nitrogen forms a triple bond.

Azanide

H_2N^-

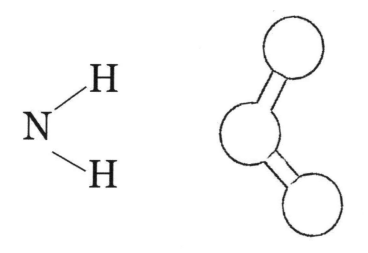

Azanide is another small compound commonly found in larger compounds. As a functional group, when paired with a carbon atom it is called an amine group. In skeletal formulas it is commonly written as either H_2N or NH_2. NH_3 and NH are also called amine groups when part of larger compounds in certain contexts.

Hydrazine

N_2H_4

$$H_2N-NH_2$$

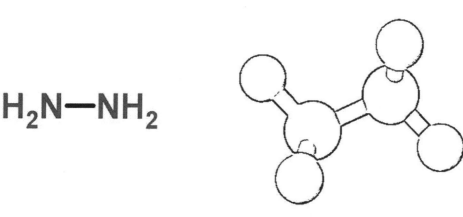

Nitrogen dioxide
NO$_2$

O=N-O

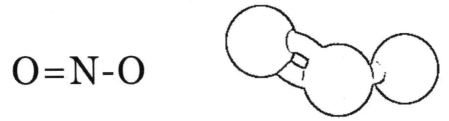

Nitrogen can also bond with two oxygen atoms. But, unlike carbon, nitrogen prefers to make only 3 bonds with other atoms. Nitrogen needs 3 electrons to fill its valence shell. Oxygen needs 2 electrons to fill its valence shell.

In the figure above, the oxygen on the right still wants an electron to fill its valence shell. What happens next is called **resonance**. Resonance describes the nature of delocalized electrons. Because both oxygen atoms want a double bond with nitrogen, they pull on the electrons equally.

O-N=O
O=N-O

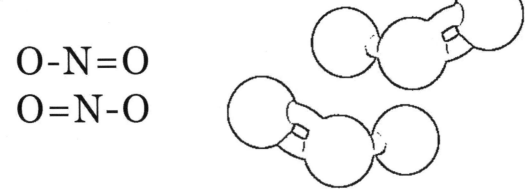

The two resonance structures of nitrogen dioxide are as shown above.

However, even the resonance structures do not give the whole picture. The actual bonds between nitrogen and the two oxygen atoms are identical to each other. Instead of thinking of them as single bonds or double bonds, each set of bonds is a blend of the two. The bonds are approximately two-thirds single bond and one-third double bond. This is difficult to convey on paper, but the below is an attempt.

O=N=O

Benzene ring
C_6H_6

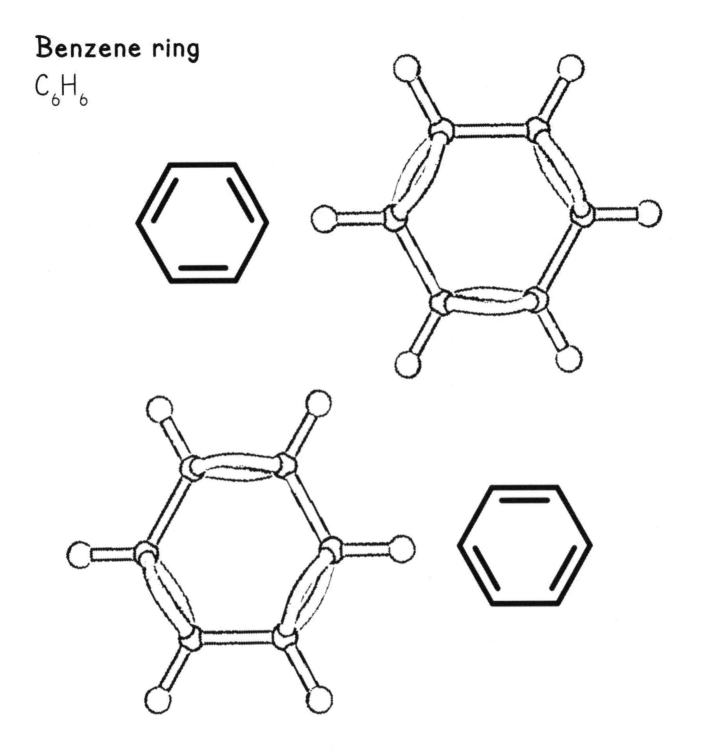

Benzene rings are found in a lot of chemical structures and are another good example of resonance.

As before, the skeletal formula does not label each carbon and hydrogen.

Benzene's skeletal formula is sometimes shown as above, with a circle in the center representing its resonant nature.

Like with nitrogen dioxide, the bonds in a benzene ring are not double bonds and single bonds, but bonds that are a sort of hybrid between the two.

Diazine

$C_4H_4N_2$

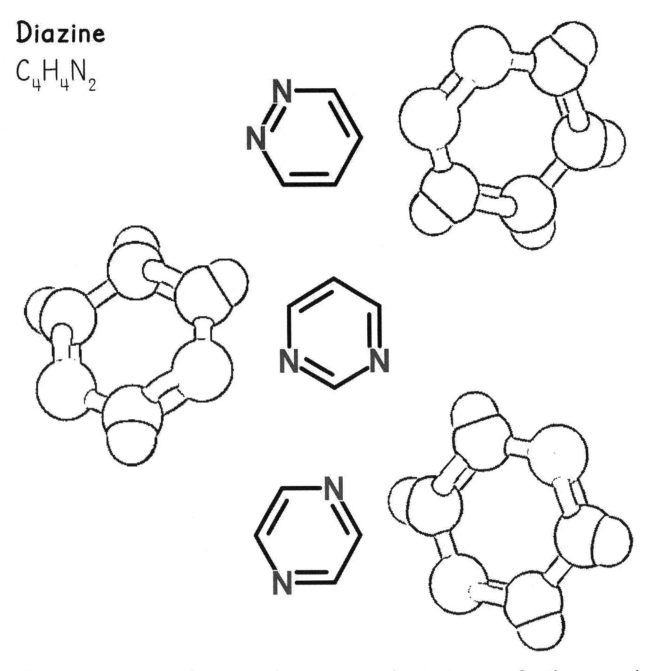

Diazine is similar to benzene, but two of the carbon atoms are replaced by nitrogen atoms. There are three forms of diazine, as shown above.

Like benzene, diazine has resonance, but for simplicity it is commonly shown as having double and single bonds. There are many different ring structures found in chemistry in addition to benzene and diazine.

Other Covalent Compounds

Phosphorus trichloride
PCl_3

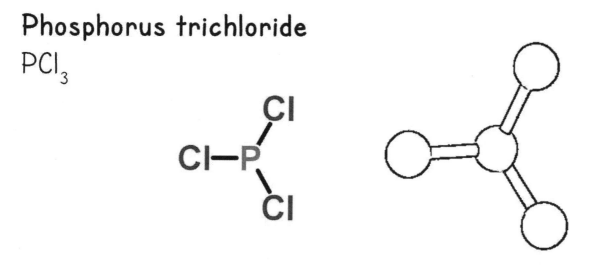

Phosphorus is in the same column as nitrogen on the periodic table. It has 5 electrons in its valence shell and likes to gain 3 electrons. Chlorine is a halogen with 7 electrons in its valence shell. It likes to gain 1 electron.

Phosphorus(V) chloride
PCl_5

Phosphorus can also bond with five chlorine atoms. It is a little complicated, but phosphorus can move some of its electrons to fill the *3d* sublevel. Covalent bonds can form in various ways, as this compound demonstrates.

Tetraphosphorus trisulfide
P_4S_3

Sulfur is in the same column on the periodic table as oxygen. It has 6 electrons in its valence shell and likes to gain 2 electrons. Phosphorus and sulfide can bond together in many different ways. One of these ways is as shown above in tetraphosphorus trisulfide.

Sulfur dibromide
SBr_2

Bromine is a halogen like chlorine. Bromine has 7 valence electrons and makes a single bond to complete its outer shell. Sulfur can bond to two bromine atoms to form sulfur dibromide.

Formaldehyde
CH$_2$O

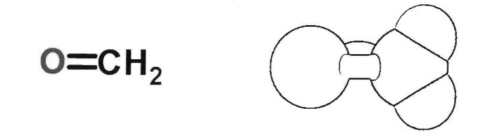

Formaldehyde is an example of a compound with a carbonyl group. A carbonyl group consists of a carbon double-bonded to an oxygen atom.

Formic acid
HCOOH

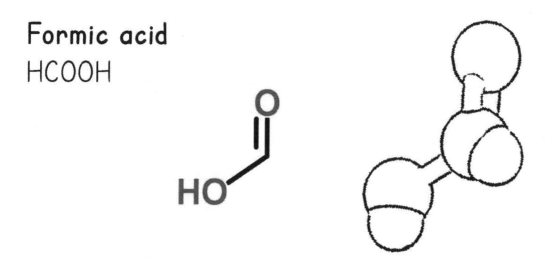

A carboxyl group is like a carbonyl group, but in addition to a double bond to an oxygen atom the carbon is also bonded to a hydroxyl group. Compounds with carboxyl groups are called carboxylic acids or organic acids.

To show how some of the previous structures you've seen can come together to form more complex compounds, examples of nucleobases, neurotransmitters, hormones and vitamins will be used. For each compound, mark an X beside the functional groups you can find in them.

Nucleobases

Nucleobases are the building blocks of our DNA and RNA. Without these, we would not be who we are.

Adenine

$C_5H_5N_5$

Adenine is found in DNA and RNA.

___amine group   ___carbonyl group   ___carboxyl group
___hydroxyl group ___methyl group ___methylene group
___ring structure

Cytosine

$C_4H_5N_3O$

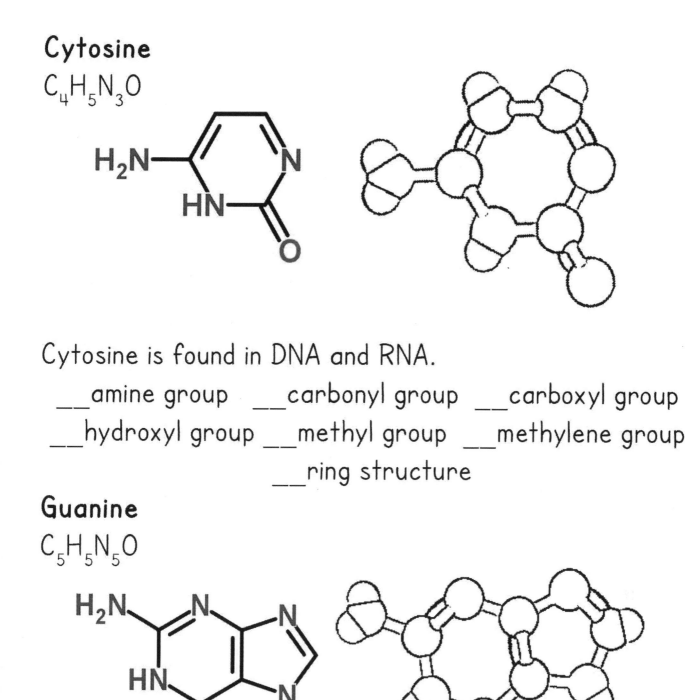

Cytosine is found in DNA and RNA.

___amine group   ___carbonyl group   ___carboxyl group
___hydroxyl group ___methyl group   ___methylene group
___ring structure

Guanine

$C_5H_5N_5O$

Guanine is found in DNA and RNA.

___amine group   ___carbonyl group   ___carboxyl group
___hydroxyl group ___methyl group   ___methylene group
___ring structure

Thymine

$C_5H_6N_2O_2$

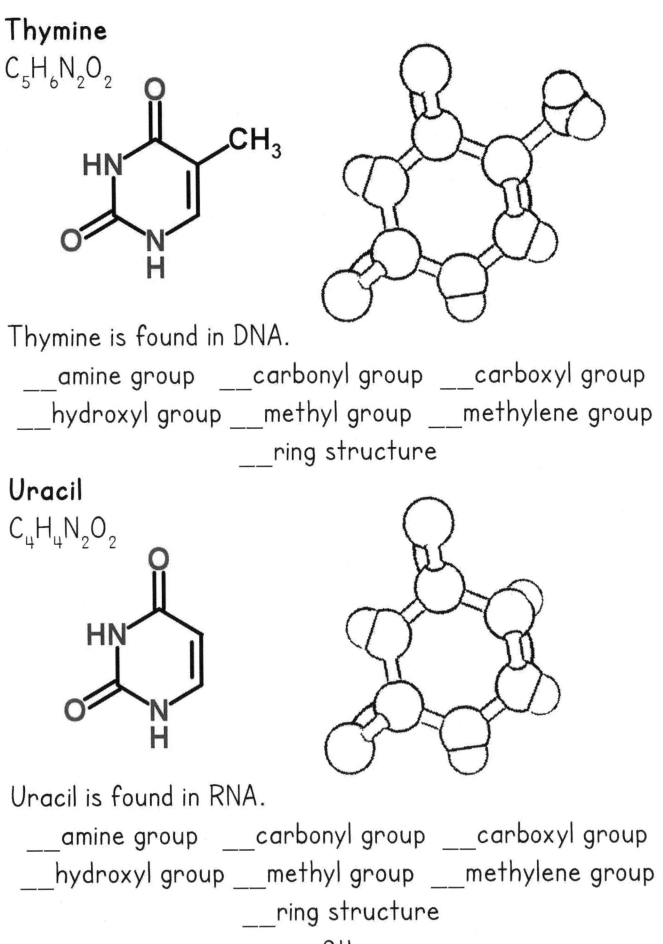

Thymine is found in DNA.

___amine group   ___carbonyl group   ___carboxyl group

___hydroxyl group ___methyl group ___methylene group

___ring structure

Uracil

$C_4H_4N_2O_2$

Uracil is found in RNA.

___amine group   ___carbonyl group   ___carboxyl group

___hydroxyl group ___methyl group ___methylene group

___ring structure

Neurotransmitters

Neurotransmitters are the chemical messengers of our nervous system. There are many neurotransmitters at work in our brains. Here are a few of them.

Adrenaline (also called epinephrine)

$C_9H_{13}NO_3$

__amine group   __carbonyl group   __carboxyl group
__hydroxyl group __methyl group __methylene group
__ring structure

Noradrenaline (also called norepinephrine)

$C_8H_{11}NO_3$

__amine group   __carbonyl group   __carboxyl group
__hydroxyl group __methyl group __methylene group
__ring structure

85

Dopamine

$C_8H_{11}NO_2$

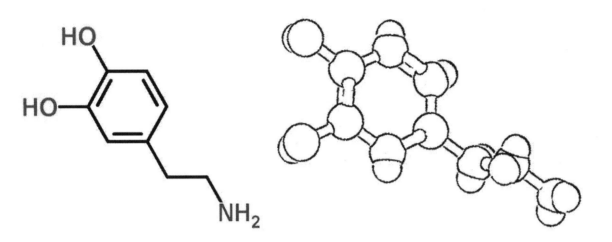

__amine group   __carbonyl group   __carboxyl group
__hydroxyl group __methyl group   __methylene group
__ring structure

Melatonin

$C_{13}H_{16}N_2O_2$

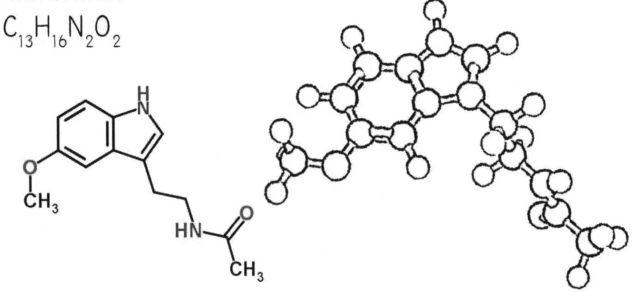

__amine group   __carbonyl group   __carboxyl group
__hydroxyl group __methyl group   __methylene group
__ring structure

Serotonin

$C_{10}H_{12}N_2O$

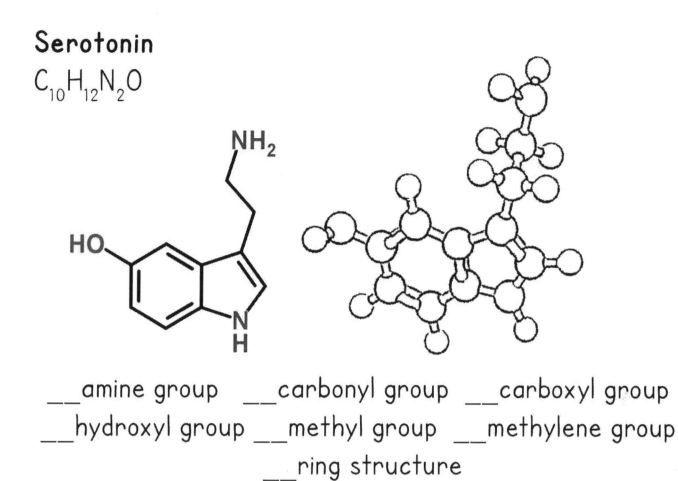

__amine group   __carbonyl group   __carboxyl group

__hydroxyl group __methyl group __methylene group

__ring structure

Hormones

Hormones are the chemical messengers of our endocrine system. There are many hormones at work in our bodies. Here are a few of them.

Cortisol

$C_{21}H_{30}O_5$

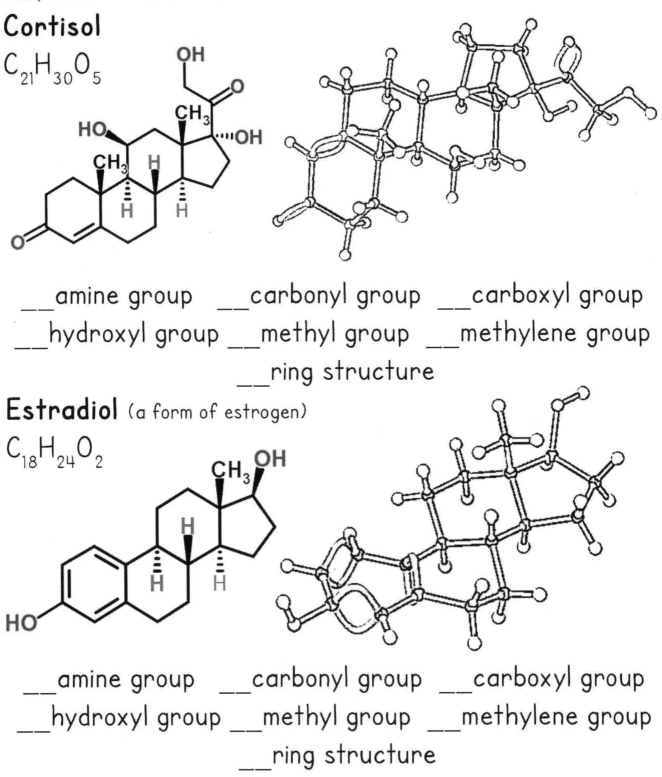

___amine group   ___carbonyl group   ___carboxyl group
___hydroxyl group ___methyl group ___methylene group
___ring structure

Estradiol (a form of estrogen)

$C_{18}H_{24}O_2$

___amine group   ___carbonyl group   ___carboxyl group
___hydroxyl group ___methyl group ___methylene group
___ring structure

Progesterone

$C_{21}H_{30}O_2$

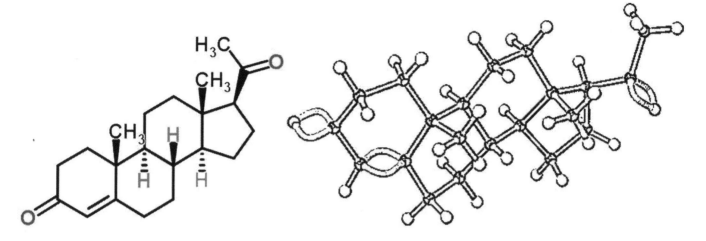

___amine group   ___carbonyl group   ___carboxyl group
___hydroxyl group ___methyl group   ___methylene group
___ring structure

Testosterone

$C_{19}H_{28}O_2$

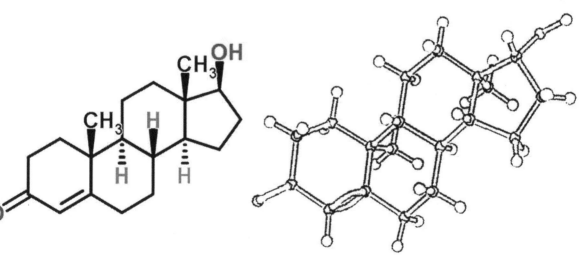

___amine group   ___carbonyl group   ___carboxyl group
___hydroxyl group ___methyl group   ___methylene group
___ring structure

Vitamins

Our body needs vitamins to function properly. Sometimes we get vitamins in our diet and sometimes our body converts other compounds into vitamins.

Retinol (Vitamin A)

$C_{20}H_{30}O$

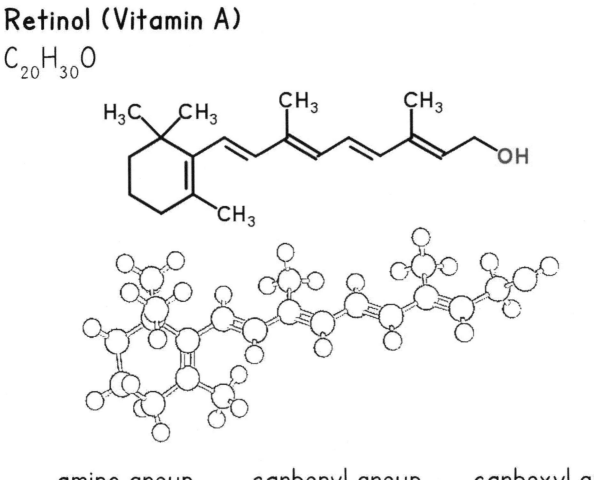

__amine group   __carbonyl group   __carboxyl group
__hydroxyl group __methyl group __methylene group
__ring structure

Beta-Carotene (the body converts beta-carotene into retinol)

$C_{40}H_{56}$

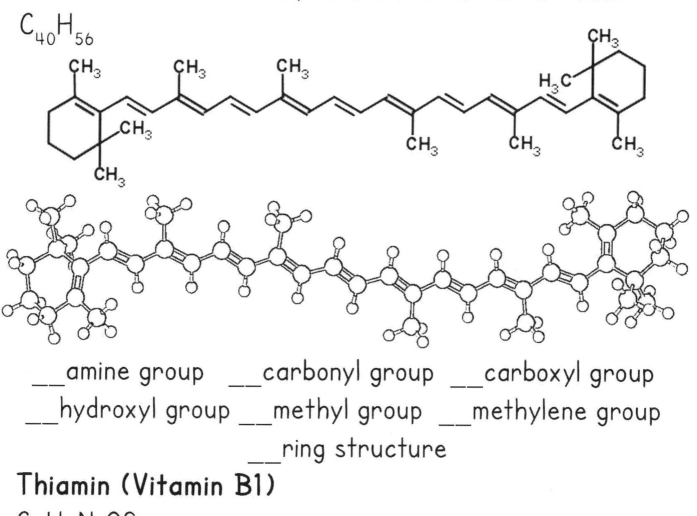

___amine group   ___carbonyl group   ___carboxyl group

___hydroxyl group ___methyl group   ___methylene group

___ring structure

Thiamin (Vitamin B1)

$C_{12}H_{17}N_4OS$

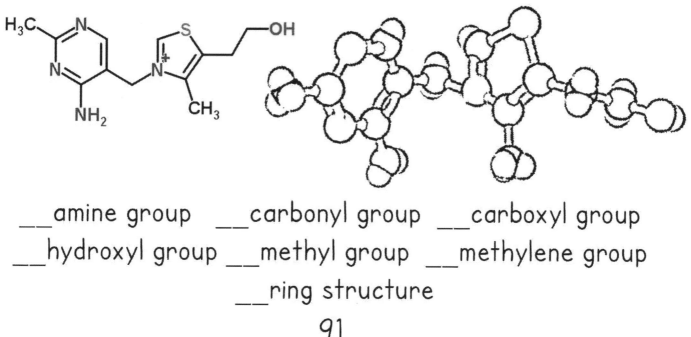

___amine group   ___carbonyl group   ___carboxyl group

___hydroxyl group ___methyl group   ___methylene group

___ring structure

Riboflavin (Vitamin B2)

$C_{17}H_{20}N_4O_6$

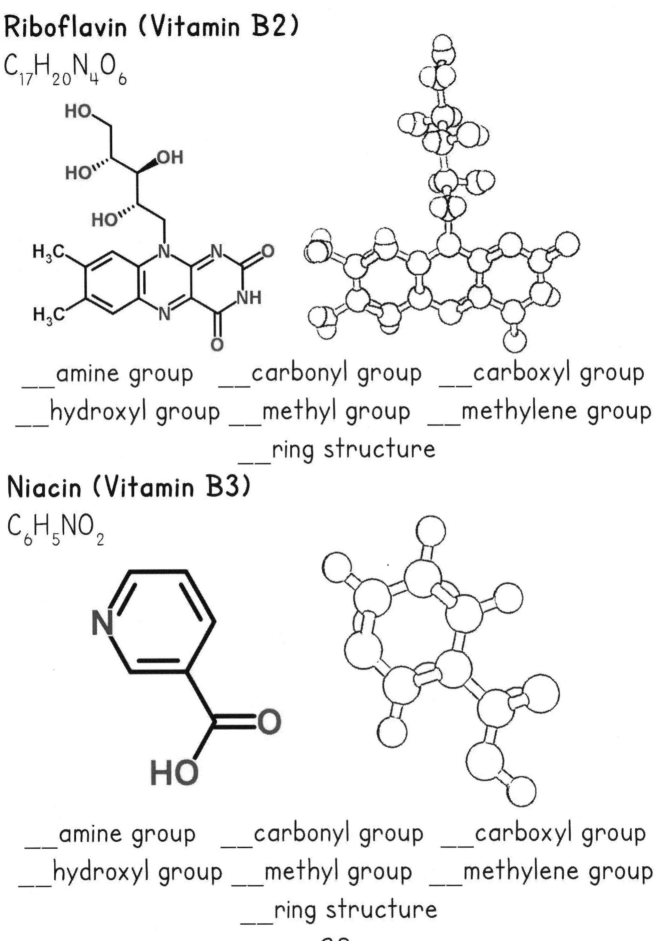

___amine group   ___carbonyl group   ___carboxyl group

___hydroxyl group ___methyl group ___methylene group

___ring structure

Niacin (Vitamin B3)

$C_6H_5NO_2$

___amine group   ___carbonyl group   ___carboxyl group

___hydroxyl group ___methyl group ___methylene group

___ring structure

92

Pantothenic Acid (Vitamin B5)

$C_9H_{17}NO_5$

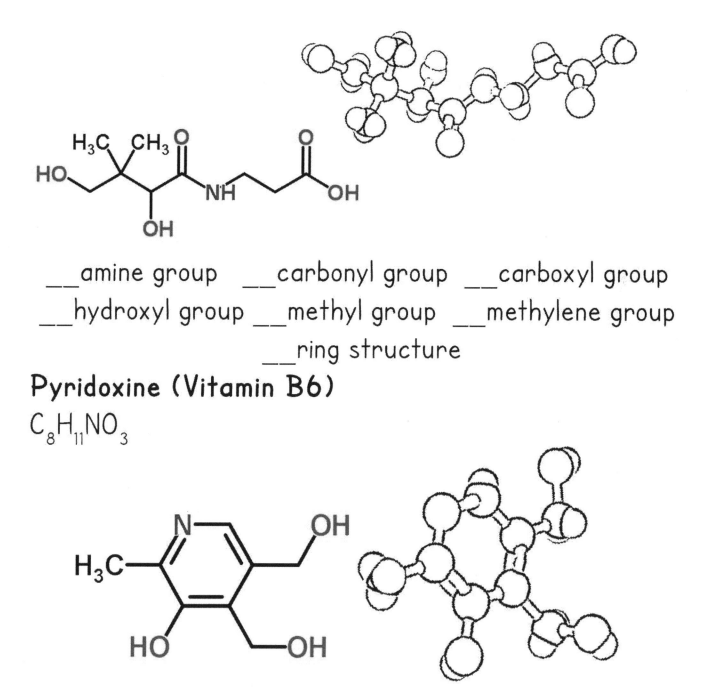

__amine group   __carbonyl group   __carboxyl group
__hydroxyl group __methyl group __methylene group
__ring structure

Pyridoxine (Vitamin B6)

$C_8H_{11}NO_3$

__amine group   __carbonyl group   __carboxyl group
__hydroxyl group __methyl group __methylene group
__ring structure

Biotin (Vitamin B7)

$C_{10}H_{16}N_2O_3S$

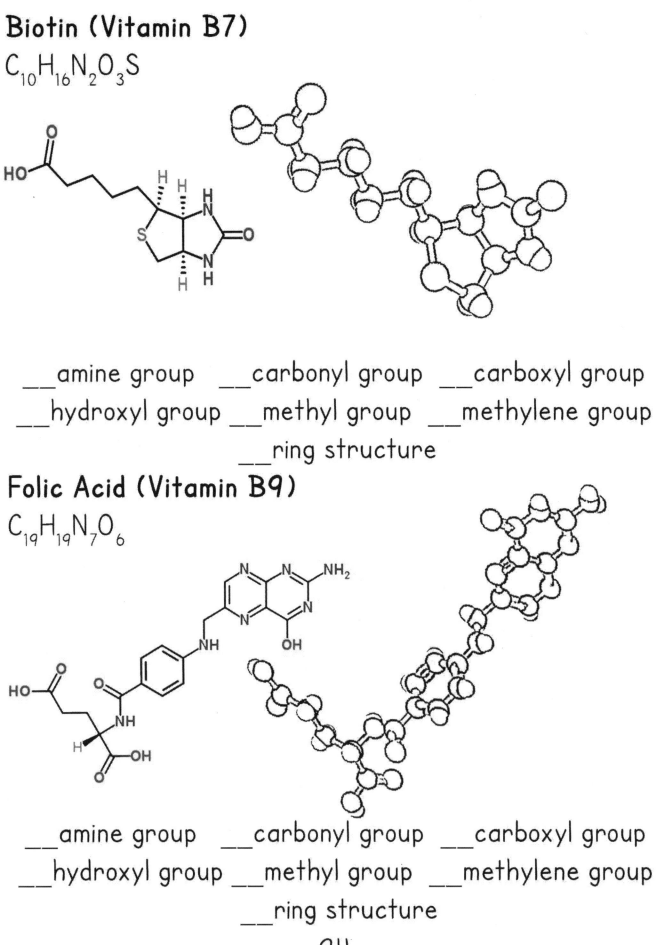

__amine group   __carbonyl group   __carboxyl group
__hydroxyl group __methyl group __methylene group
__ring structure

Folic Acid (Vitamin B9)

$C_{19}H_{19}N_7O_6$

__amine group   __carbonyl group   __carboxyl group
__hydroxyl group __methyl group __methylene group
__ring structure

Cyanocobalamin (Vitamin B12)

$C_{63}H_{88}CoN_{14}O_{14}P$

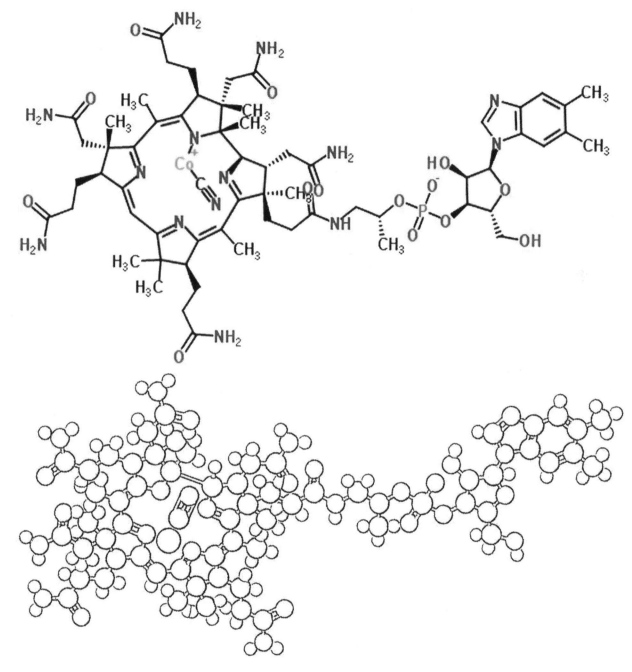

___amine group   ___carbonyl group   ___carboxyl group

___hydroxyl group ___methyl group ___methylene group

___ring structure

Ascorbic Acid (Vitamin C)

$C_6H_8O_6$

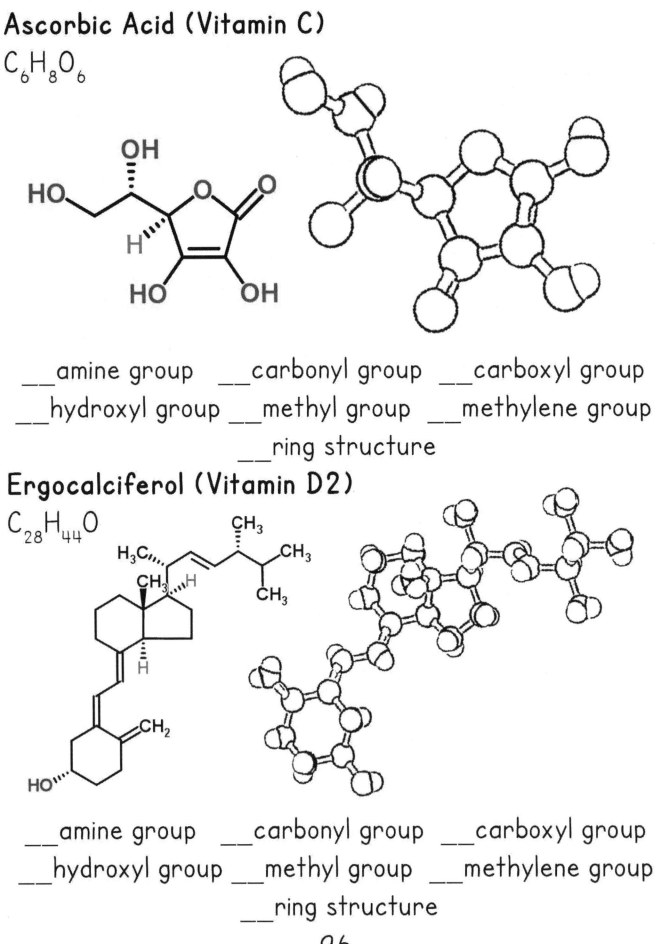

__amine group   __carbonyl group   __carboxyl group

__hydroxyl group __methyl group   __methylene group

__ring structure

Ergocalciferol (Vitamin D2)

$C_{28}H_{44}O$

__amine group   __carbonyl group   __carboxyl group

__hydroxyl group __methyl group   __methylene group

__ring structure

Cholecalciferol (Vitamin D3)

$C_{27}H_{44}O$

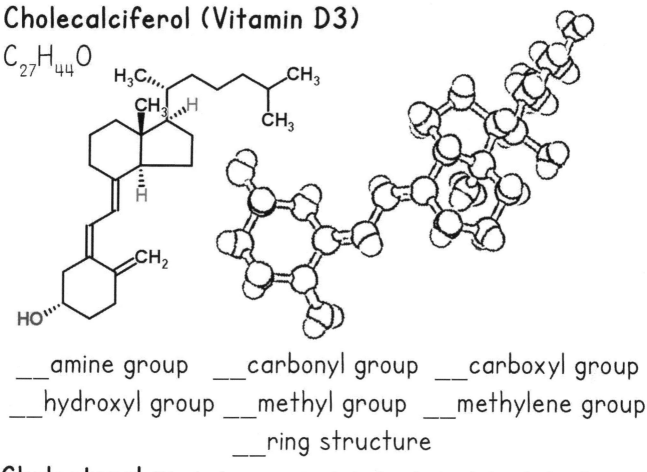

__amine group __carbonyl group __carboxyl group

__hydroxyl group __methyl group __methylene group

__ring structure

Cholesterol (the body converts cholesterol into cholecalciferol)

$C_{28}H_{46}O$

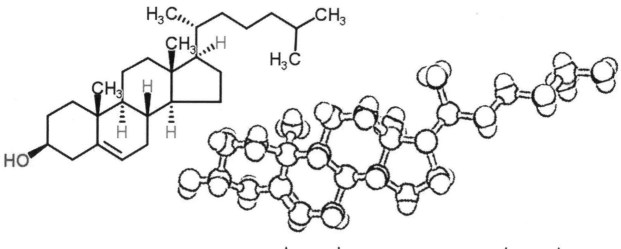

__amine group __carbonyl group __carboxyl group

__hydroxyl group __methyl group __methylene group

__ring structure

Alpha-tocopherol (Vitamin E)

$C_{29}H_{50}O_2$

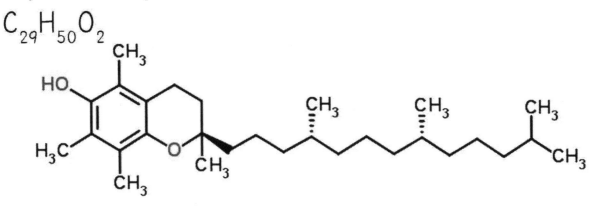

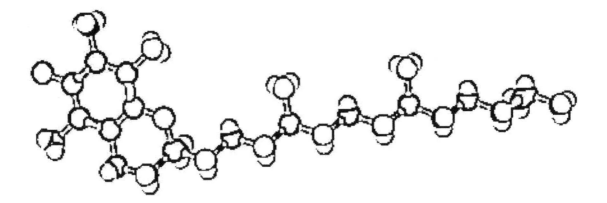

__amine group   __carbonyl group   __carboxyl group
__hydroxyl group __methyl group __methylene group
__ring structure

Phytonadione (Vitamin K1)

$C_{31}H_{46}O_2$

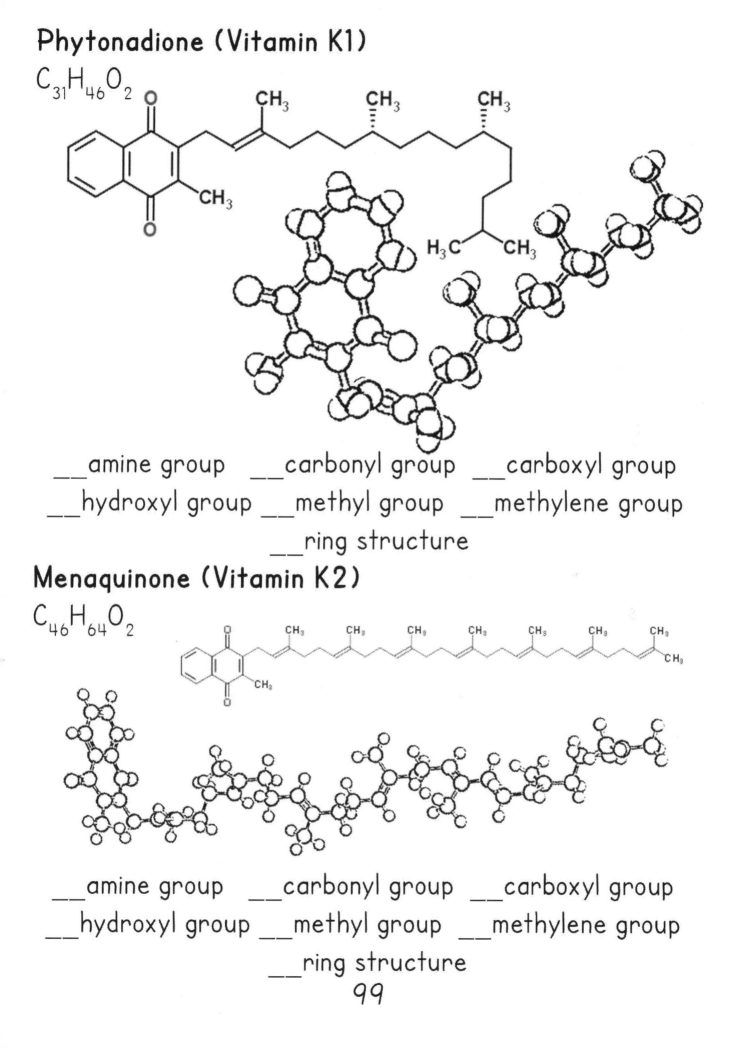

CH₃ CH₃ CH₃
CH₃
H₃C CH₃

__amine group   __carbonyl group   __carboxyl group

__hydroxyl group __methyl group __methylene group

__ring structure

Menaquinone (Vitamin K2)

$C_{46}H_{64}O_2$

CH₃ CH₃ CH₃ CH₃ CH₃ CH₃ CH₃
CH₃
CH₃

__amine group   __carbonyl group   __carboxyl group

__hydroxyl group __methyl group __methylene group

__ring structure

99

Common Polyatomic Ions

Polyatomic ions are compounds with either a negative or positive charge due to gained or missing electrons. Try to remember them and their charges.

Ammonium
NH_4^+

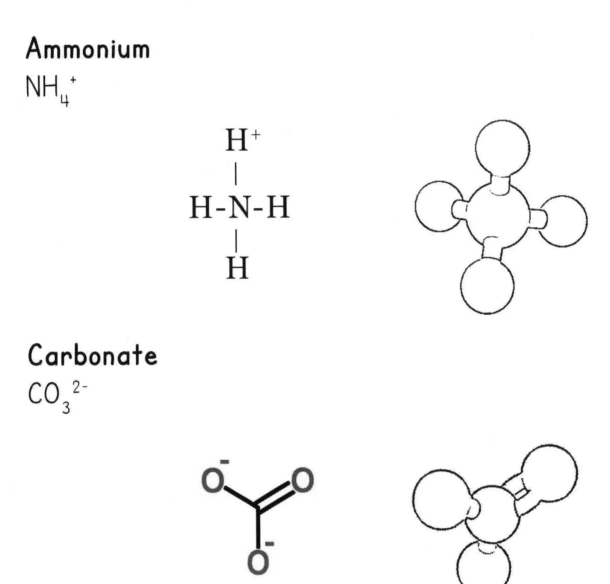

Carbonate
CO_3^{2-}

Bicarbonate

HCO_3^-

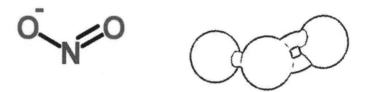

Nitrite

NO_2^-

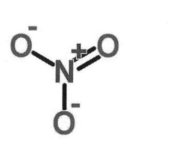

Notice that nitrite looks the same as nitrogen dioxide (page 74). The two are composed of the same atoms, but nitrite has gained an extra electron while nitrogen dioxide has a balanced number of protons and electrons.

Nitrate

NO_3^-

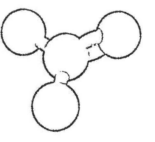

Chlorite
ClO⁻

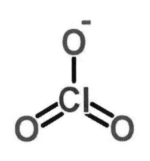
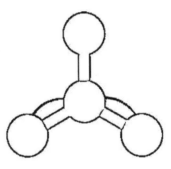

Chlorate
ClO₃⁻

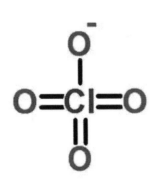
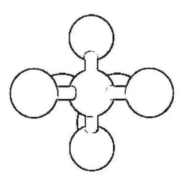

Perchlorate
ClO₄⁻

Sulfite

SO_3^{2-}

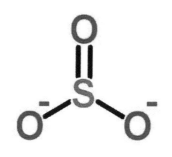
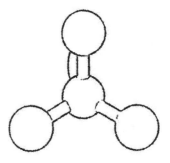

Sulfate

SO_4^{2-}

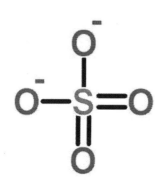
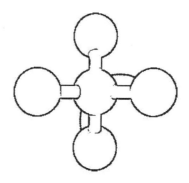

Thiosulfate

$S_2O_3^{2-}$

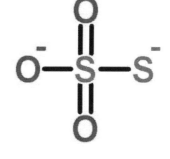
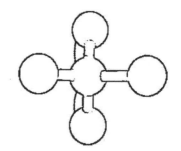

Phosphite

PO_3^{3-}

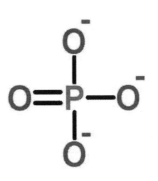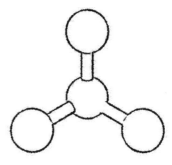

Phosphate

PO_4^{3-}

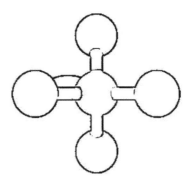

More Ionic Compounds

You've seen how metals bond with atoms of nonmetals. Metals bond with polyatomic ions the same way. Bonds form so the charge is balanced.

Sodium bicarbonate
$NaHCO_3$

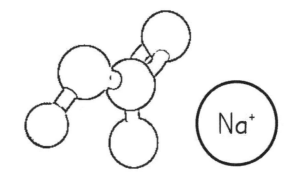

Sodium sulfite
Na_2SO_3

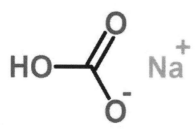

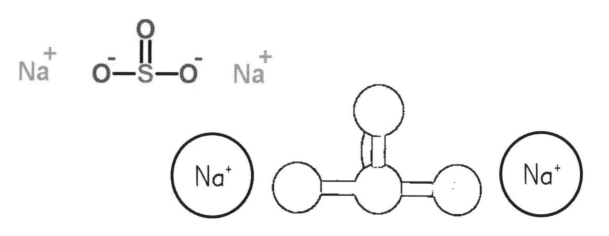

Calcium Bicarbonate
Ca(HCO$_3$)$_2$

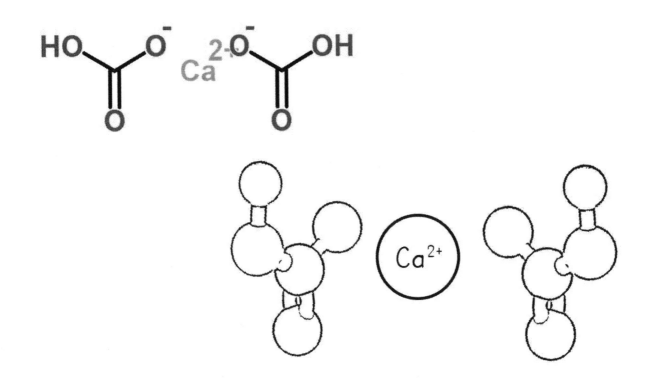

Calcium Sulfite
CaSO$_3$

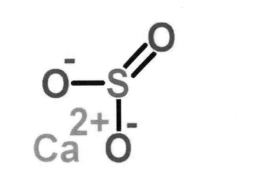

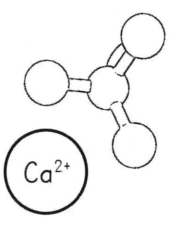

Polyatomic ions with opposite charges can form compounds too.

Ammonium sulfate
$(NH_4)_2SO_4$

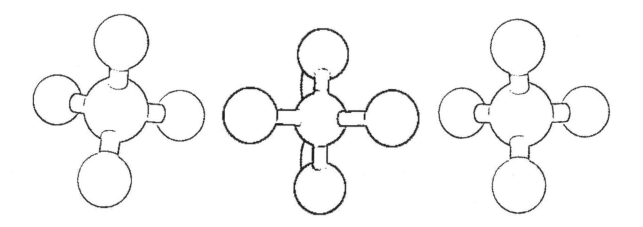

Memory Exercises

It is good to memorize the atomic symbols of the elements. See if you can memorize these elements' symbols and then test yourself on the following pages.

H	Hydrogen	K	Potassium	Xe	Xeon
He	Helium	Ca	Calcium	Cs	Cesium
Li	Lithium	Cr	Chromium	Ba	Barium
Be	Beryllium	Mn	Manganese	W	Tungsten
B	Boron	Fe	Iron	Pt	Platinum
C	Carbon	Co	Cobalt	Au	Gold
N	Nitrogen	Ni	Nickel	Hg	Mercury
O	Oxygen	Cu	Copper	Pb	Lead
F	Fluorine	Zn	Zinc	Po	Polonium
Ne	Neon	As	Arsenic	At	Astatine
Na	Sodium	Se	Selenium	Rn	Radon
Mg	Magnesium	Br	Bromine	Fr	Francium
Al	Aluminum	Kr	Krypton	Bh	Bohrium
Si	Silicon	Rb	Rubidium	Cn	Copernicum
P	Phosphorus	Sr	Strontium	Es	Einsteinium
S	Sulfur	Ag	Silver	Bi	Bismuth
Cl	Chlorine	Sn	Tin	Zr	Zirconium
Ar	Argon	I	Iodine	Tl	Thalium

It is also good to memorize the formulas and charges of common ions and polyatomic ions. Try to memorize these and then test yourself on the following pages.

H^+	Hydrogen ion	ClO^-	Chlorite
CO_3^{2-}	Carbonate	Na^+	Sodium ion
Mg^{2+}	Magnesium ion	ClO_4^-	Perchlorate
SO_3^{2-}	Sulfite	NO_3^-	Nitrate
Li^+	Lithium ion	PO_3^{3-}	Phosphite
Cl^-	Chlorine ion	K^+	Potassium ion
PO_4^{3-}	Phosphate	$S_2O_3^{2-}$	Thiosulfate
I^-	Iodine ion	NH_4^+	Ammonium
HCO_3^-	Bicarbonate	Cs^+	Cesium ion
O^{2-}	Oxygen ion	Br^-	Bromine ion
F^-	Fluorine ion	Be^{2+}	Beryllium ion
SO_4^{2-}	Sulfate	Al^{3+}	Aluminum ion
N^{3-}	Nitrogen ion	Sr^{2+}	Strontium ion
ClO_3^-	Chlorate	Fr^+	Francium ion
Ca^{2+}	Calcium ion	Rb^+	Rubidium ion
Ba^{2+}	Barium ion	Ra^{2+}	Radium ion
NO_2^-	Nitrite		

_____ Hydrogen	_____ Potassium	_____ Xeon
_____ Helium	_____ Calcium	_____ Cesium
_____ Lithium	_____ Chromium	_____ Barium
_____ Beryllium	_____ Manganese	_____ Tungsten
_____ Boron	_____ Iron	_____ Platinum
_____ Carbon	_____ Cobalt	_____ Gold
_____ Nitrogen	_____ Nickel	_____ Mercury
_____ Oxygen	_____ Copper	_____ Lead
_____ Fluorine	_____ Zinc	_____ Polonium
_____ Neon	_____ Selenium	_____ Astatine
_____ Sodium	_____ Arsenic	_____ Radon
_____ Magnesium	_____ Bromine	_____ Francium
_____ Aluminum	_____ Krypton	_____ Bohrium
_____ Silicon	_____ Rubidium	_____ Copernicum
_____ Phosphorus	_____ Strontium	_____ Einsteinium
_____ Sulfur	_____ Silver	_____ Bismuth
_____ Chlorine	_____ Tin	_____ Zirconium
_____ Argon	_____ Iodine	_____ Thalium

_____ Hydrogen ion

_____ Carbonate

_____ Magnesium ion

_____ Sulfite

_____ Lithium ion

_____ Chlorine ion

_____ Phosphate

_____ Iodine ion

_____ Bicarbonate

_____ Oxygen ion

_____ Fluorine ion

_____ Sulfate

_____ Nitrogen ion

_____ Chlorate

_____ Calcium ion

_____ Barium ion

_____ Nitrite

_____ Chlorite

_____ Sodium ion

_____ Perchlorate

_____ Nitrate

_____ Phosphite

_____ Potassium ion

_____ Thiosulfate

_____ Ammonium

_____ Cesium ion

_____ Bromine ion

_____ Beryllium ion

_____ Aluminum ion

_____ Strontium ion

_____ Francium ion

_____ Rubidium ion

_____ Radium ion

Hydrogen	Potassium	Xeon
Helium	Calcium	Cesium
Lithium	Chromium	Barium
Beryllium	Manganese	Tungsten
Boron	Iron	Platinum
Carbon	Cobalt	Gold
Nitrogen	Nickel	Mercury
Oxygen	Copper	Lead
Fluorine	Zinc	Polonium
Neon	Selenium	Astatine
Sodium	Arsenic	Radon
Magnesium	Bromine	Francium
Aluminum	Krypton	Bohrium
Silicon	Rubidium	Copernicum
Phosphorus	Strontium	Einsteinium
Sulfur	Silver	Bismuth
Chlorine	Tin	Zirconium
Argon	Iodine	Thalium

_____ Hydrogen ion	_____ Chlorite
_____ Carbonate	_____ Sodium ion
_____ Magnesium ion	_____ Perchlorate
_____ Sulfite	_____ Nitrate
_____ Lithium ion	_____ Phosphite
_____ Chlorine ion	_____ Potassium ion
_____ Phosphate	_____ Thiosulfate
_____ Iodine ion	_____ Ammonium
_____ Bicarbonate	_____ Cesium ion
_____ Oxygen ion	_____ Bromine ion
_____ Fluorine ion	_____ Beryllium ion
_____ Sulfate	_____ Aluminum ion
_____ Nitrogen ion	_____ Strontium ion
_____ Chlorate	_____ Francium ion
_____ Calcium ion	_____ Rubidium ion
_____ Barium ion	_____ Radium ion
_____ Nitrite	

	Hydrogen		Potassium		Xeon
	Helium		Calcium		Cesium
	Lithium		Chromium		Barium
	Beryllium		Manganese		Tungsten
	Boron		Iron		Platinum
	Carbon		Cobalt		Gold
	Nitrogen		Nickel		Mercury
	Oxygen		Copper		Lead
	Fluorine		Zinc		Polonium
	Neon		Selenium		Astatine
	Sodium		Arsenic		Radon
	Magnesium		Bromine		Francium
	Aluminum		Krypton		Bohrium
	Silicon		Rubidium		Copernicum
	Phosphorus		Strontium		Einsteinium
	Sulfur		Silver		Bismuth
	Chlorine		Tin		Zirconium
	Argon		Iodine		Thalium

_____ Hydrogen ion	_____ Chlorite
_____ Carbonate	_____ Sodium ion
_____ Magnesium ion	_____ Perchlorate
_____ Sulfite	_____ Nitrate
_____ Lithium ion	_____ Phosphite
_____ Chlorine ion	_____ Potassium ion
_____ Phosphate	_____ Thiosulfate
_____ Iodine ion	_____ Ammonium
_____ Bicarbonate	_____ Cesium ion
_____ Oxygen ion	_____ Bromine ion
_____ Fluorine ion	_____ Beryllium ion
_____ Sulfate	_____ Aluminum ion
_____ Nitrogen ion	_____ Strontium ion
_____ Chlorate	_____ Francium ion
_____ Calcium ion	_____ Rubidium ion
_____ Barium ion	_____ Radium ion
_____ Nitrite	

Notes

Notes

Notes

Notes

Notes

Notes

Notes